History and GIS

Alexander von Lünen • Charles Travis
Editors

History and GIS

Epistemologies, Considerations and Reflections

Springer

Editors
Alexander von Lünen
Department of Geography
University of Portsmouth
United Kingdom

Charles Travis
Trinity College
University of Dublin
Ireland

ISBN 978-94-007-9829-8 ISBN 978-94-007-5009-8 (eBook)
DOI 10.1007/978-94-007-5009-8
Springer Dordrecht Heidelberg New York London

Printed on acid-free paper

Springer is part of Springer Science+Business Media (www.springer.com)

Preface

Historical geographical information systems (GIS) have seen a considerable rise in popularity over recent years in some historical sub-disciplines, particularly in economic and social history, for example. If one examines the programs of social science history conferences, such as the annual meetings of the Social Science History Association (SSHA) or the bi-annual European Social Science History Conference (ESSHC), for instance, historical GIS are featured very prominently.

However, these disciplines are but fringe groups in history departments across the world, albeit vociferous ones. If one looks at the major national historical organizations (e.g., the Royal Historical Society, the American Historical Association, the Deutsche Historikertag), the reaction most likely to be encountered by HGIS advocates borders on bemusement and bewilderment. In spite of historians seeing the attractiveness of GIS (and digital tools in general), the majority of "mainstream" historians do not seem to be willing to engage with the technology. It may be attributed to the wider lack of quantitative methods in history curricula these days—and some HGIS practitioners are arguing exactly this point—that computers are rarely used by historians other than for word processing and literature databases. This in turn begs the question: is the only thing that digital tools have to offer historians are computerized quantitative methodologies? Surely not, as there are a good number of tools now being used by social scientists to handle qualitative data. Putting aside that qualitative GIS has been discussed by geographers, the question for historians remains: why bother?

Perhaps the question should be "why not"? The recent emergence of the digital humanities as a viable disciplinary path brings this question front and center. If historians are going to engage the "where" of the events that they study, considering the benefits and limitations of GIS is certainly appropriate. Critical GIS practitioners in geography have recently discussed—despite social theory critiques of its Euclidian, positivistic, techno-centric focus—ways of thinking outside of the conventional "GIS Box." Recent ethnographic and feminist engagements as well as uses of GIS to study literature and to illustrate critical spatial theory are conversant with tropes engaged by historians.

With the "spatial turn" in full swing in the humanities and many historians dealing with spatial and geographical questions, one would think GIS would be welcomed with open arms. Yet there seems to be no general anticipation by historians of employing GIS as a research tool. As mentioned, HGIS are popular chiefly among historical geographers and social and economic historians. Other historical sub-disciplines, such as ancient history, are also very open to this emerging technology since the scarcity of written sources in this field can be mitigated by inferences made from an HGIS that has archaeological data stored in it, for example. In most of modern history, however, the use of GIS is rarely seen. The intellectual benefit that a GIS may bring about seems not apparent to scholars from this subdiscipline (and others).

At the end of the day, GIS poses the daunting challenge of a steep learning curve for those who are not too savvy in geography/cartography, database technology and statistical analysis. Why would historians enter this technical maze when they cannot expect to find something worthwhile for their research agenda at its exit? Let us not forget that quantitative methods (including thematic cartography) had their halcyon days in the 1960s—both in geography and history! Since then, many—if not most—scholars from both subjects (and in other disciplines as well) have departed from this approach not because number crunching proved to be too intellectually challenging, but to be too intellectually and theoretically fruitless. Quantitative methods did not cease to be popular because people became ignorant, but because there emerged a sense that statistics would not answer the essential questions in their respective disciplines.

To a great extent, the use of GIS has been limited in its approach, and this is perhaps the main reason why it is lagging behind other fields in terms of methodology and—ultimately—in terms of recognition and acceptance. Geographical-spatial issues are experiencing heretofore unseen popularity, be it in the form of Google Earth used in the evening news or GPS navigation systems in cars, be it the metaphorical "spatial turn" in the humanities—space and geography are the talk of the day. But one has to wonder whether this is a breakthrough or just the latest fad—or both. Given the current popularity of spatial topics, GIS should be the killer app[1] for digital history. Yet, the majority of historians, though curious, remain reluctant to engage in GIS. Is the reason for this merely technical?

We do not think so. The reason for GIS being so little used in mainstream history is because GIS and its proponents have so little to offer in terms of intellectual merit, but are asking for so much in terms of learning curve. Any historian trying to get into GIS is not only bombarded with technical challenges that are way over their head; he or she has also gotten the blame laid firmly at his or her doorstep by GIS aficionados for this inaptitude. While the technical competence of many historians might be an issue, and one could hope for a change from the younger generation which has grown up with digital technology, one cannot help but to have

[1]In IT lingo, a "killer application" refers to an application that furthers the breakthrough of an underlying technology. The electronic word processor and spread sheet programs, for example, can be regarded as the killer applications for PCs in the 1980s.

sympathy with those historians remaining unconvinced, given that many HGIS are about digitization rather than analysis and that those projects that analyze data do so in a way that has been kicked in the long grass sometime ago by most historians (such as quantitative methods).

After some years of discomfort at conferences, seminars and other academic venues, we believe that the time is right to discuss and consider methods rather than projects. All too often, books, articles, and conference sessions have concerned themselves with technical issues of GIS in history (such as how to handle temporal data), and not with the intellectual benefit the non-initiated could hope to gain. We therefore have asked authors across a wide range of academic disciplines to share their experiences and thoughts about why historians should use GIS and not how.

The opening chapter is by *David Bodenhamer*, director of The Polis Center, a research unit of the IU School of Liberal Arts at Indiana University–Purdue University Indianapolis (USA). Bodenhamer is a well-established scholar in the field of historical GIS, having published extensively on the topic, such as the volume The Spatial Humanities: GIS and the Future of Humanities Scholarship which he co-edited with Trevor Harris and was published in 2010. In his chapter in this book, Bodenhamer outlines the general issues with GIS historians might encounter.

Next is an interview conducted by co-editor *Alexander von Lünen* with famous French historian *Emmanuel Le Roy Ladurie* of the renowned *Annales* school. Le Roy Ladurie was one of the first historians to use computers and geography in historical research and talks about his life and work, which sheds some light on the current issues in history and GIS.

Historian *Onno Boonstra* from the Radboud Universiteit in Nijmegen (the Netherlands) reports on the history of choropleth mapping in the Netherlands to predict the headway GIS may make in historical scholarship. Boonstra has been working with GIS for some years now but remains skeptical about its ability to leave a lasting mark in history departments.

Geographer *Mark H. Palmer* from the University of Missouri (USA) works on social aspects of GIS and conducts research on Native American history. Rather than speaking about the methodology of GIS in general historical research, he details the issues that arise when GIS is used to map historical accounts from cultures that have a non-Western approach to both geography and history. In his chapter, Palmer delves into postcolonial and sociological discourses to point out how researching and writing the history of indigenous people can and should be done with GIS.

Historians *Detlev Mares* from the Technische Universität Darmstadt (TUD, Germany) and *Wolfgang Moschek* from the Private Tagesheim- und Internatsschule Dieburg (Germany) look at the value of GIS in the teaching of history, both at high-school and university level. Mares is a senior principal lecturer in the history department at the TUD and—among other things—is responsible for didactic and educational issues in the department (such as developing courses), and Moschek is a high-school teacher for history and geography. They look at the way that GIS might help to shape and develop a "spatial historical competence" in students.

Renowned Swedish philosopher of geography *Gunnar Olsson* is interviewed by co-editor *Alexander von Lünen* next. Olsson is on a life-long quest to define (and criticize) the nature of cartographical reasoning, that is, the knowledge that we can gain from maps. He gives some very inspiring thoughts for the historical GIS community, in his discussion of the "geographical inference problem" (i.e. what inferences about the world can we make from geographical information).

Historian of science *Alexi Baker* from the History and Philosophy of Science department of the University of Cambridge (UK) then describes her motivation to get involved with GIS technology while doing her Ph.D. work a few years ago at Oxford. Not having a technical background, she outlines her approach to GIS and how it helped her analyzing early modern geographies.

Medieval historian *Faye Taylor* from the Cumberland Lodge (a charitable educational institution) in Nottingham (UK) then gives an overview how GIS might be used to help researching medieval religious history (the topic of her recently finished PhD at the University of Nottingham) in mapping religious apparitions in a GIS and relating this to research methods in medieval history. All pre-1800 history is confronted by the issue of very inaccurate and or simply non-available geographical information, and Taylor surveys projects in medieval history that have used GIS.

Computer scientist and geographer *Monica Wachowicz* from the Geodesy and Geomatics Engineering department at the University of New Brunswick (Canada) and historian *J.B. "Jack" Owens* from Idaho State University (USA), in their chapter explain the latter's concept of "geographically-integrated history" and how this relates to so-called Knowledge Spaces in order to shift from quantitative methods in HGIS to more qualitative ones. The authors propose a new kind of HGIS, drawing on methods from knowledge representation, data mining, natural language processing, machine learning, information retrieval, and visual analytics.

Following is an interview with historian *David J. Staley* conducted by co-editor *Charles Travis*. Staley has done extensive work on the role of visualization in history, including visualization done with GIS, and is a pioneer in digital history. He gives his thoughts on GIS and what role it might play in historical scholarship.

Urban historian *Sam Griffiths* from the Faculty of the Built Environment at University College London (UK) looks at some general issues that historians may have with the spatial concept of historical GIS. Working on the boundaries between history, geography, and architecture, Griffiths points out that a lot of intellectual debate is necessary to arrive at a concept of space that makes GIS applicable to "mainstream" historians.

Geographer *Charles B. Travis*, a research associate with the Trinity Long Room Hub at the Trinity College, University of Dublin (Ireland), then contextualizes GIS within the tradition of geographical thought and practice, before attempting a translation of the technology as a means to write critical three-dimensional spatial histories.

Edward L. Ayers, *Robert K. Nelson*, and *C. Scott Nesbit* from the Digital Scholarship Lab, University of Richmond (USA), give an account of the American Historical Atlas from 1932 to look at an early fusion of history and geography and how this informs us about the role GIS might play in contemporary historical research.

Computer scientist and historian *Alexander von Lünen* from the Geography department at the University of Portsmouth (UK) finally looks to relate GIS to actual historical scholarship by detailing general methodological issues and how GIS might fit in there. Rather than to make the historian come to the GIS, he argues, we have to make the GIS come to the historian—not by developing new HGIS, but by (ab)using existing GIS technology to harness it in a way GISystems and GIScience would not think it possible (or permissible, for that matter).

Portsmouth, UK Alexander von Lünen
Dublin, Ireland Charles Travis

Acknowledgements

Part of Alexander von Lünen's work on this book was funded by the Leverhulme Trust under research project grant A20090049: "Data Models for Actor-Networks in Historical GIS: Workers and localities in C19."

Charles Travis would like to thank Professor Emeritus Anne Buttimer, Professor David Nemeth, Professor Patrick Duffy, Dr. David Drew, and Dr. Mark Hennessy for giving him the guidance and space so he could develop his idiosyncratic GIS approaches.

The editors would also like to thank the Royal Geographical Society (RGS) for hosting their session "There is no place like home—why historians would want to use a GIS" (which was the point of departure for this book) at the RGS annual meeting 2010 (and in particular the History and Philosophy of Geography Research Group of the RGS for sponsoring us at that event) and the Association of American Geographers (AAG) for hosting the session "There is no place like home—why historians would want to use a GIS, part 2" at the AAG annual meeting 2011, where the decision was made to publish this book.

Many thanks also to Dr. Robert Doe at Springer for helping us making this book becoming a reality.

Last, but not least, thanks to our wives, partners, friends, and families for their love and support.

Contents

Chapter 1
Beyond GIS: Geospatial Technologies and the Future of History

David J. Bodenhamer

1.1 The Spatial Turn

Academic disciplines periodically undergo reorientation as core themes shift to re-direct research and focus attention on new questions or reexamine traditional problems from new perspectives. In the humanities and social sciences, a recent line of inquiry has focused on space, prompting scholars of society and culture to talk about a spatial turn within their disciplines.[1] Even a cursory review of literature reveals the influence of this new direction. Subject matter, once organized largely as periods and eras, increasingly is ordered under spatial themes, such as region, diaspora, contact zones, and borders or boundaries, among others. This shift has been accompanied by and reinforced through an equivalent concern with material culture, built and natural environment, and other markers of space and place.

It is not the first time that attention to space and time has reshaped the way we approach social and cultural questions. A similar turn occurred from 1880 to 1920 when a series of sweeping technological changes created new ways of thinking about time and space. Distance-collapsing innovations—the telephone, wireless telegraph, radio, cinema, automobiles, and airplanes, among others—challenged traditional understandings of how time and space intersected with the social world. It suddenly was possible to know events as they occurred, and this experience of simultaneity refashioned people's sense of distance and direction. It also meant that individuals were no longer cut off from the flow of time; widely available film and photographic images made the past as accessible as the present, while new developments in science and the world fairs that showcased them made the future

[1] A number of recent titles explore this spatial turn in the humanities, including Bodenhamer et al. (2010), Dear et al. (2011), and Daniels et al. (2011). Also see Doorn (2005).

D.J. Bodenhamer (✉)
The Polis Center, Indiana University – Purdue University Indianapolis, Indianapolis, IN, USA
e-mail: intu100@iupui.edu

A. von Lünen and C. Travis (eds.), *History and GIS: Epistemologies, Considerations and Reflections*, DOI 10.1007/978-94-007-5009-8_1,

seem more definite and real. New scientific theories, business practices, and cultural forms reinforced the shift: Einstein's theory of relativity and Freud's conception of psychoanalysis shaped consciousness directly; time-management studies, such as Taylorism, dominated manufacturing; and James Joyce and Marcel Proust explored how to link time and space in novels, while the Cubists challenged notions of spatial perspective and form that had long dominated art.[2]

A continuous thread links the first spatial turn with the one we have experienced more recently, but it is likely that this second turn will have a more profound influence on the theory and practice of history and the humanities. The early twentieth-century conceptions of space and time had less effect on the study of the past than it did on art and literature. Frederick Jackson Turner's frontier thesis and its emphasis on the development of the American West and the history it spawned were exceptions, as was the decade-long work of the Annales movement[3]; both schools reflected an intentional focus on questions of space and time. But the cataclysms of the mid-twentieth century, from world wars and revolutions to mass movements for equality, ultimately spurred historians to search for the roots of momentous events in ideas and politics and technological or social change, causes for which spatial markers were less pronounced. The considerations for space did not disappear, but they became marked by particularity, an emphasis on place, as scholars began to discern how the story of change differed from one location to another. The focus on place reflected and reinforced a postmodernist unease with the grand narrative, which created a literature that increasingly became fragmented, with analyses existing at different geographical and temporal scales and few efforts made to link them. For many humanists, space itself became less geographical and more metaphorical, as scholars found richer meaning in conceptual space—for instance, gendered space, racialized space, or the body as space—than in categories related to the physical environment, the traditional frame of definition for spatial terms.[4]

Today, historians and other humanists are acutely aware of the social and political construction of space and its particular expression as place. Spaces are not simply the setting for historical action but are a significant product and determinant of change. They are not passive settings but the medium for the development of culture: "space is not an empty dimension along which social groupings become structured," sociologist Anthony Giddens notes, "but has to be considered in terms of its involvement in the constitution of systems of interaction" (Giddens 1984, 364). All spaces contain embedded stories based on what has happened there. These stories are both individual and collective, and each of them links geography (space) and history (time). More important, they all reflect the values and cultural codes present in the various political and social arrangements that provide structure to society. In this sense, then, the meaning of space, especially as place or landscape, is always being constructed through the various contests that occur over power.

---

[2]For more on this earlier spatial turn, see Kern (1983).

[3]See the interview with Emmanuel Le Roy Ladurie in this book to learn more about the *Annales*.

[4]Cresswell (2004) offers a good brief introduction to the postmodern construction of place. Also see Haulttunen (2006), Biernacki and Jordan (2002).

There is nothing new in this development—the earliest maps reveal the power arrangements of past societies—but humanities scholarship increasingly reflects what may in fact, by the greatest legacy of postmodernism, the acknowledgment that our understanding of the world itself is socially constructed.[5]

At its core, the spatial turn rejects the universal truths, grand narratives, and structural explanations that dominated the social sciences and the humanities during much of last century. Above all, it is about the particular and the local, without any supposition that one form of culture is better than another. Its claim is straightforward: To understand human society and culture, we must understand how it developed in certain circumstances and in certain times and at certain places. From this knowledge, we can appreciate that the world is not flat but incredibly complicated and diverse. This view no longer seems new because humanists have embraced it eagerly; now we all recognize the particularity of space, the importance of place. But for all the uses we make of this insight—and for all its explanatory power—the concepts of space and place employed by historians frequently are metaphorical and not geographical. Far less often have we grappled with how the physical world has shaped us or how in turn we have shaped perceptions of our material environment.

1.2 GIS and History

New spatial technologies, especially geographic information systems (GIS), are now facilitating a (re)discovery of geographical space in history and the other humanities. At its core, GIS is a powerful software that uses location to integrate and visualize information. Within a GIS, users can discover relationships that make a complex world more immediately understandable by visually detecting spatial patterns that remain hidden in texts and tables. Maps have served this function for a long time, but GIS brings impressive computing power to this task. Its core strength is its ability to integrate, analyze, and make visual a vast array of data from different formats, all by virtue of their shared geography. This capability has attracted considerable interest from historians, archaeologists, linguists, students of material culture, and others who are interested in place, dense coil of memory, artifact, and experience that exists in a particular space, as well as in the coincidence and movements of people, goods, and ideas that have occurred across time in spaces large and small. Recent years have witnessed a wide-ranging, if still limited, application of GIS to historical and cultural questions: Did the Dust Bowl of the 1920s and 1930s result from over-farming the land or was it primarily the consequence of larger term environmental changes? What influence did the rapidly changing cityscape of London have on literature in Elizabethan England? What is the relationship between

[5]Michel de Certeau reminds us that "space occurs as the effect produced by the operations that orient it, situate it, temporalize it, and make it function as a polyvalent unity of conflictual programs or contractual proximities." And stories are the constructive means we use to transform spaces into places or places into spaces. See de Certeau (1984, 117–118).

rulers and territory in the checkered political landscape of state formation in the nineteenth-century Germany? How did spatial networks influence the administrative geography of medieval China? What spatial influences shaped the development of the transcontinental railroad in the United States? Increasingly, scholars have turned to GIS to provide new perspective on these and other topics that previously have been studied outside of an explicitly spatial framework.[6]

Despite this flurry of interest and activity, most uses to date of GIS in historical and cultural studies have been disparate, application driven, and often tied to the somewhat its more obvious role in census boundary delineation and map making. While not seeking to minimize the importance of such work, these studies have rarely addressed the broader, more fundamental issues that surround the introduction of a spatial technology such as GIS into the humanities. There are core reasons why GIS has found early use and ready acceptance in the sciences and social sciences rather than in the more qualitatively based humanities. The humanities pose far greater epistemological and ontological issues that challenge the technology in a number of ways, from the imprecision and uncertainty of data to concepts of relative space, the use of time as an organizing principle, and the mutually constitutive relationship between time and space. Essentially, GIS and its related technologies currently allow users to determine a geometry of space. In the context of the humanities, it will be necessary to move GIS from this more limited quantitative representation of space to facilitate an understanding of place within time and the role that place occupies in humanities disciplines.

In their essence, historians seek to generalize from the particular, not for the purpose of finding universal laws but rather to glean insights about cause and effect from a known outcome. Here, the humanities differ from much of social science, which attempts to reach a generalization that holds true in any similar circumstance. This difference is significant and influences the way the two groups of scholars create knowledge. For many social scientists, the search for trustworthy generalization focuses on the isolation of an independent variable, the cause that has a predictable effect on dependent variables or ones that respond to the stimulus or presence of a catalyst. They believe it is possible to discover such a variable, given sufficient resources, because the world is not yet lost to them. Historians must contend with fragmentary evidence and are painfully aware that the past is incomplete and irretrievable. They also are skeptical of prediction. The past cannot be changed, even if its interpretation as history is continually in flux, but in it the intersection of patterns and singular events can be discovered. Not so the future, where continuities and contingencies coexist independently of one another.

Historians view reality as weblike, to use philosopher Michael Oakeshott's phrase, because they see everything as related in some way to everything else.

---

[6]See Knowles (2008a) for a good sample of the application of GIS to various topics in the humanities. Also see the special issue of *The International Journal of Humanities and Arts Computing*, vol 3, no. 1–2, 2009, which is devoted to the use of GIS in a number of humanities disciplines.

Interdependency is the lingua franca of the humanities, and most recently, it has become embodied in practice theory, or in the view of one of its leading proponents, historian William Sewell, "social life may be conceptualized as being composed of countless happenings or encounters in which persons or groups of persons engage in social action." In this view, societies and social systems are "continually shaped and reshaped by the creativity and stubbornness of their human creators" (Sewell 2005, 110–111). Another historian, Ed Ayers, has labeled this concept "deep contingency," an effort to understand society as a whole with "all structures put into motion and motion put into structures" (Ayers 2010, 7).[7]

The goal of historical scholarship is not to model or replicate the past; a model implies the working out of dependent and independent variables for purposes of prediction, whereas replication suggests the ability to know the past and its cultural forms more completely than most humanists would acknowledge is possible. Humanists practice an extractive scholarship: they have the capacity for selectivity and shifting perspectives in the pursuit of the fullest possible understanding of heritage and culture. Traditionally, humanities scholars have used narrative to construct the portrait that furthers this objective. Narrative encourages the interweaving of evidentiary threads and permits the scholar to qualify, highlight, or subdue any thread or set of them—to use emphasis, nuance, and other literary devices to achieve the complex construction of past worlds. All of these elements—interdependency, narrative, and nuance, among others—predispose the humanists to look askance at any method or tool that appears to reduce complex events to simple schemes. An insistence on precision does not fit the worldview of humanities scholars; indeed, these disciplines appear at times to embrace an uncertainty principle—the more precisely you measure one variable, the less precise are other variables.

It is no accident that historians have embraced eclectic methods as fervently as they resist anything that smacks of reductionism. Questions drive historical scholarship, not hypotheses, and the questions that matter most address causation: "why" matters more than "whom," "what," or "when," even though these latter questions are neither trivial nor easy to answer. The research goal is not to eliminate explanations or to disprove the hypothesis but to open the inquiry through whatever means are available and by whatever evidence may be found. This sense of eclectic borrowing has long informed humanities scholarship and even finds strong advocates among some of the most well-known theorists in the humanities, hence the advice offered by Paul de Mans to the Irvine Critical Literature department about its mission to develop "a new kind of skill [...] the capacity to use and feel at home in a whole series of different critical and theoretical codes and systems, as one would use a particular foreign language, without remaining rigidly locked into any one of them, but rather developing the capacity to translate those findings into different codes, systems, critical positions, as the case may require."[8]

[7] See also Ayers et al.'s contribution in this book.

[8] http://chronicle.com/blogPost/how-theory-damaged-the-humanities/6178, last accessed 27 Oct 2011.

A well-presented argument often does not settle a question; it may complicate it or open new questions that previously were unimagined. Similarly, historians are hard-pressed to identify a preferred method because each avenue of investigation yields different evidence and thus different insights. Historians revisit evidence as they discover new data. Their approach is recursive, not linear: The goal is not so much to eliminate answers as to admit new perspectives. These methods doubtless appear quixotic to non-humanists because they do not lead to finality. But for humanists, the goal is not proof but meaning. The challenge, then, is to use geospatial technologies to probe, explore, challenge, and complicate—in sum, to allow us to see, experience, and understand human behavior in all its complexity and to view its deep contingency. As in traditional humanities scholarship, the aim is less to produce an authoritative or ultimate answer than to prompt new questions, develop new perspectives, and advance new arguments or interpretations.

Seeking to fuse GIS with history and the humanities is challenging in the extreme. Latent tension, if not direct conflict, exists in linking a positivist technology with predominantly humanist traditions. The epistemological clash is most apparent in the emphasis within traditional GIS on quantitative data, precise measurement, coordinate systems, and spatial models. Categorizing geographic complexity into entities, fields, objects, attributes, and geometric topology contrasts starkly with the humanities emphasis on ambiguity, complexity, nuance, and plurality. The scientific method that underpins GIS with its computational demands for accuracy and precision, a Euclidian coordinate space, and its algorithmic emphasis on generalization and reductionism stands in sharp relief with the humanist emphasis on the individual and the unique, on contingency and emergent realities, and on narrative as a way to weave the complex threads of space, time, and artifact. It is misleading to speak simply of a qualitative-oriented humanism and a quantitative GIS, yet it also is evident that sharp differences exist between the conceptual mapping of the humanities and the cartographic mapping of GIS. At its core, GIS privileges disambiguation in its organization of knowledge, whereas the humanities treat knowledge as multivalent, equivocal, and protean.

Data and the representations of phenomena, then, are singular factors that challenge the fusion of GIS with the humanities. Yet the GIS abstractions of space, nature, and society, while posing substantial problems, are particularly relevant in the humanities where notions and representations of place, rather than those of space, are primary. To this end, GIScientists have made recent advances in spatial multimedia, in GIS-enabled web services, geovisualization, cybergeography, exploratory spatial data analysis, and virtual reality that provide capabilities far exceeding the abilities of GIS on its own. Together, these technologies have the potential to revolutionize the role of place in the humanities by moving beyond the two-dimensional map to explore dynamic representations and interactive systems that will prompt an experiential, as well as rational, knowledge base.

This notion of a richer, dynamic, and experiential GIS resonates with the evocative and thick descriptions of place and time that humanists have long favored in their scholarship. Even mapping itself comports well with the aims and methods of humanists. Representation of the past, suggests historian John Lewis Gaddis, is a kind of mapping where the past is a landscape and history is the way we fashion it. The metaphor, one consistent with disciplinary traditions across the humanities, makes the link between "pattern recognition as the primary form of human perception and the fact that all history [...] draws upon the recognition of such patterns" (Gaddis 2002, 33). In this sense, mapping is not cartographic but conceptual. It permits varying levels of detail, not just as a reflection of scale but also of what is known at the time. Like the map, history becomes better and more accurate as we continue to accumulate more detail, observe its patterns, and refine our knowledge.

Significantly, the discipline that provided the home for much GIS development and application, geography, found itself divided over the technology in ways that mimicked the concerns expressed by humanists about quantitative methods generally. The central issue was, at heart, epistemological: GIS privileges a certain way of knowing the world, one that values authority, definition, and certainty over complexity, ambiguity, multiplicity, and contingency, the very things that engaged humanists. From this internal debate came efforts to reposition GIS as GIScience, a shift that corresponded with other efforts to embody the technology with a theoretical framework that it previously lacked. This intellectual restructuring pushed the technology in new directions that were more suitable to the humanities.[9]

Ultimately, what will compel attention from historians and other humanities scholars are the broader ontological and epistemological issues of geographic information science (GISci) and not GIS as a method and technique. The juncture of GISci with the humanities generates a more fertile and intellectually rewarding basis for conceptualizing and representing space than does the spatial tool kit of GIS. GIS is not a panacea for the humanities; its appropriate use demands good judgment and a broader knowledge of the production of space than the simple application of a technology can provide. The spatial turn in the humanities must be more than method: it must understand the role of space in human events. To that end, the spatial humanities need to embrace more than geospatial technologies but also geographical concepts of space.

There are more methods and approaches available to scholars to explore the spatial humanities than the very heavy emphasis on off-the-shelf-software packages provided by GIS vendors. To date, the spatial turn in the humanities is predominantly a GIS-enabled rediscovery of the power of the map, yet many historians have used the 2D format and defaults provided by the software even though these methods, in addition to being cartographically uninformed, flatten the world the humanist seeks to understand. Better suited to the humanities is the related field of geovisualization, ranging from dynamic maps to virtual reality and cybergeography. It offers insightful new ways of seeing and understanding spatial relationships. It also provides significant potential for integrating qualitative data in the form of images and text, the primary evidence used by historians. Virtual GIS combines the elements of virtual reality and serious gaming, with the

---

[9]Natalie Schuurman relates this development in Schuurman (2004, 21–52); also see Sheppard (2005).

spatial analytic and data-handling capability of GIS to provide an immersive and experiential environment that mimics what historians often seek in their written descriptions of the past. These innovative geovisualization methods, and the spatial analysis they embody, not only avoid the subject-object paradigm embedded in traditional mapping, but they also provide a more powerful framework for using all the available evidence and not simply data in its quantitative form.[10]

Within the field of cultural heritage, archaeologists have used GIS and computer animations to reconstruct the Roman Forum, for example, creating a 3-D world that allows users to walk through buildings that no longer exist, except as ruins. We can experience these spaces at various times of the day and seasons of the year. We see more clearly a structure's mass and how it clustered with other forms to mold a dense urban space. In this virtual environment, we gain an immediate, intuitive feel for proximity and power. This constructed memory of a lost space helps us recapture a sense of place that informs and enriches our understanding of ancient Rome (Digital Roman Forum Project).[11] A similar, although more ambitious, project uses laser-scanning technology (LiDAR, Light Detection and Ranging) to create 3-D models of major heritage sites and allows scholars and others to roam this virtual environment at will.[12] In similar fashion, historians and material culturists have joined with archeologists to fashion Virtual Jamestown.[13] This project, in turn, is seedbed for an even more ambitious attempt to push the technology toward the humanities by placing Jamestown at one vertex of Atlantic world encounters. Its goal is to repopulate a virtual world with the sense of possibilities embedded in the past, what Paul Carter has called "intentional history" (Carter 1987, 3). Viewed within the spatial context for their actions, which includes the presence of proximate cultures, whether indigenous tribes, Spanish, Africans, or Dutch, we then can understand better how contingencies became lost as they butted against the encountered realities within the space the English claimed in 1607.

Historians often fail to see the potential of these approaches for addressing scholarly problems, viewing them instead as mere reconstructions or as artificial environments far removed from the world of causation and argument. What we fail to realize is how much even simple virtual geographic environments can prompt the sympathetic imagination that is at the core of good scholarship. Consider Gettysburg, the most-studied battle of the American Civil War. Historians have used a wide array of evidence to provide minute-by-minute descriptions of the battle yet often have failed to analyze carefully what the generals actually could

---

[10] A good brief introduction to the various forms of virtual geographic environments—3D models and 2.5D extruded surfaces, computer animations, interactive models, virtual globes, online virtual worlds, games, and semi- or fully immersive virtual reality—can be found in Priestnall et al. (2012).

[11] http://dlib.etc.ucla.edu/projects/Forum/, last accessed on 1 Aug 2008

[12] http://www.ted.com/talks/ben_kacyra_ancient_wonders_captured_in_3d.html, last accessed 3 Jan 2012

[13] http://www.virtualjamestown.org, last accessed on 14 Aug 2008

see (as opposed to what they reported as seeing), even though visual awareness and visual reconnaissance were essential elements in nineteenth-century warfare. A GIS-facilitated reconstruction of the battleground provided clarity on the sightlines and viewsheds enjoyed by commanders, thereby enhancing but not upsetting traditional interpretations. More importantly, the virtual environment raised new questions about issues ranging from military decisions to the emotional and psychological experiences of battle, which now could be seen and understood within the geographical environment in which they occurred (cf. Knowles 2008b). Virtual environments are only one way that GIScience is moving us beyond conventional uses of GIS and mere mapping. The Geospatial Web also holds considerable potential for the spatial humanities to automate the identification and mapping of people, events, places, and spatial relationships from textual resources. Similarly, the ability to transform unstructured text into structured maps through computational text and data mining, semantic synthesis, geoparsing, place-name matching using natural language processing and digital gazetteers, and georeferencing methods promises to be a major contribution to the armory of tools available to the spatial humanities. Spatialization techniques such as self-organizing maps and text clouds identify clusters in text documents that share similar characteristics in metaphorical space and not just geographical space. Text-to-map transformations reflect both absolute and relative space by extracting spatial relationships embedded in text and then using this information to go beyond the strict cartographic map making that dominates current humanities use of GIS (cf. Harris et al. 2010).

These developments from GIScience are important because they quickly are moving us beyond the constraints of a technology illsuited for much humanities research. The same cannot be said with as much assurance about time, which has long been the central lens through which historians view change. The spatial and temporal turns (the New Historicism) go together, and it is unwise to separate the two or prioritize one over the other. GIS has struggled to adequately handle the complexities of these spatiotemporal needs, with the result that the software emphasizes space and treats time as categorical and discontinuous. But the spatial humanities require both time and space: to speak of history as dealing with time and geography with space is too simplistic a divide. Doreen Massey's idea of exploring multiple trajectories through space and time[14] is much more suited to humanities research, but GIS struggles to provide an environment in which this integrated space-time can be explored profitably. The value of animated maps has been acknowledged in helping to understand movement as a basic characteristic of human existence, but, as with text mining, the humanities pose significant challenges to the GIS community to develop components linked specifically to their needs.

For historians and humanists generally, place is more important than space. Although place and place-making are part of the spatial in all social science and humanities disciplines, the highly structured database schemas of GIS cannot

[14]Massey (2005, 9–15) and Massey (1999). A useful discussion also may be found in Peuquet (2002).

accommodate easily, if at all, the contested dynamics of place. The technology cannot speak to the contingent nature of cultural processes or to the agents of change and transformation, nor does it handle the humanities scholar's penchant for dismembering, rethinking, and recombining. GIS also has difficulties managing deep contingency, the notion that all social life is implicated and unpredictable, nor is it good at thick description, the heavily layered ethnography that represents an effort to capture the complicated realities of social life. These methods probe the deeply fused connections of time and space that link the public, private, economic, social, political, religious, and civil realms at various scales and discover how structural transformation at one scale often results in a rupture of social processes at other scales. The focus here is less on causation than on interpreting the consequences and the resonances of events as evidenced through the intersection of place, time, and action. This deeply layered interpretive history draws heavily on space and place as an organizing framework to understand the world, yet it challenges GIS as to how this might be achieved.

1.3 The Spatial Humanities: Spatial Narratives and Deep Maps

Central to the emergence of the spatial humanities is a trust that the contingent, unpredictable, and ironic in history and culture can be embodied within a narrative context that incorporates space alongside of time. For the humanities—and for social scientists who are influenced by the humanities—it is above all the thick weave of events, locations, behaviors, and motivations that make human experience of space into place. Place is the product of deep contingency and of the human effort to render that experience meaningful in language, art, ritual, and in other ways. Place is constructed out of the imagination as much as through what is visible and tangible in experience. Humanists, social scientists, and geographers, and all who are interested in seeing a spatial humanities mature, should plan for increasingly more complex maps (using the term broadly) of the personalities, emotions, values, and poetics, the visible and invisible aspects of a place. The spatial considerations remain the same, which is to say that geographic location, boundary, and landscape remain crucial, whether we are investigating a continental landmass or a lecture hall. What must be added is a spatial narrative that acknowledges how engaged human agents build spatially framed identities and aspirations out of imagination and memory that complement the verbal narrative traditionally employed by humanists.

At its core, a spatial narrative focuses on spatial patterns as a means of understanding social interaction. It is a geography of the constant interaction between structure and process, a continuous interplay between society and the individual and/or group within a spatial environment that both shapes and is shaped by social norms and individual or group agency. As a construct within the humanities, this narrative also must accommodate time and contingency; the social interactions influenced by and influencing space represent, in fact, a web of choices, and the narrative becomes a braided thread (or multiple threads) of those choices over time. But the real question is not the definition of a spatial narrative—and these definitions are as numerous as they are abstract—but how to tap digital and spatial technologies to move narrative beyond the linear constraints of written language and disciplinary argument into a more fluid and reflexive process in which we can see and experience interaction as a way of understanding it more fully. If the current scholarly interest in networks is understood as an initial foray into the analysis of important interactions, the spatial narrative can be envisioned as a much richer and complex presentation, one that is geared to the analysis of vast data sets and undertaken in such a way as to maximize experimentation with evidence of contingency, contradiction, and oscillation in interactions over time.

Here is where the deep map becomes important. Stemming from the affective stance of unitary urbanism and psychogeography associated with the Situationists International in 1950s France, the approach "attempts to record and represent the grain and patina of place through juxtapositions and interpenetrations of the historical and the contemporary, the political and the poetic, the discursive and the sensual [...]" (Pearson and Shanks 2001, 64–65).[15] The idea of deep mapping has a counterpart in geography in the work of Yi Fu Tuan's *Topophilia* (Tuan 1974), who proposed exploring the connectedness and ties between human emotion and the physical fabric of landscape. As a new creative space, deep maps have several qualities well suited to a fresh conceptualization of GIS and other spatial technologies as they are applied to the humanities. They are meant to be visual, time-based, and structurally open. They are genuinely multimedia and multilayered. They do not seek authority or objectivity but involve negotiation between insiders and outsiders, experts and contributors, over what is represented and how. Framed as a conversation and not a statement, deep maps are inherently unstable, continually unfolding and changing in response to new data, new perspectives, and new insights.

The analogue between a deep map and advanced spatial technologies seems evident. Traditional geographic information systems operate as a series of layers, each representing a different theme and tied to a specific location on planet Earth. These layers are transparent, although the user can make any layer or combination of layers opaque while leaving others visible. A deep map of heritage and culture, centered on memory and place, ideally would work in a similar fashion. The layers of a deep map need not be restricted to a known or discoverable documentary record but could be opened, wiki-like, to anyone with a memory or artifact to contribute. However structured, these layers would operate as do other layers within a GIS, viewed individually or collectively as a whole or within groups, but all tied to time and space, which provide perspectives on the places that interest us. It is an open, visual, and experiential space, immersing users in a virtual world in which uncertainty, ambiguity, and contingency are everpresent, but all are capable of being braided into a narrative that reveals the ways in which space and time influences and

[15] Also see Bodenhamer (2008).

is influenced by social interaction. This space is one in which both horizontal and vertical movement is possible, with the horizontal providing the linear progression we associate with rational argument and vertical movement providing the depth, texture, tension, and resonance of experience.[16]

The concepts of deep contingency and deep mapping go beyond traditional uses of GIS and point to new realms for pursuing phenomenology and for representing emotion and experience within geospatial technologies. They seek to capture the essence of place and a humanistic sense of distance, direction, and identity. Deep mapping moves the user from the GIS world of observation to one of habitation where the material world is experienced through our own embodiment and sense of "being in the world." Nonrepresentational theory and the concepts of deep contingency, deep mapping, taskscapes and affordances, and thick description enable scholars to engage the material world rather than observe it and interrelate theories of practice and agency and how people both create their material world and, in turn, are created by it. This linking of critical geographies, postmodern humanities, and GIScience creates a fresh conceptualization of a humanities-friendly GIS.

What is required ultimately is not simply the integration of GIS into the spatial humanities but of geography and geographical concepts as well. When this occurs—and the increasingly rich conceptual frameworks and tool kits of Web 2.0 and 3.0 suggest the time is not distant—we will fulfill the promise of a spatial humanities that draws on a GIS-enabled fusion of qualitative and quantitative data, that focuses on both space and place, that acknowledges the reality of space-time, that is multi-scalar and dynamic, and that enables the dense layering and deep mapping of place. Grounded in experiential as well as objective space, it will provide a representation of society and culture, past and present, with all its rich contradictions and complexities. It will, above all, be a conceptual and technological framework that is sensitive to the needs of historians and humanities scholars.

References

Ayers EL (2010) Turning toward space, place, and time. In: Bodenhamer DJ, Corrigan J, Harris TM (eds) The spatial humanities: GIS and the future of humanities scholarship. Indiana University Press, Bloomington

Biernacki R, Jordan J (2002) The place of space in the study of the social. In: Joyce P (ed) The social in question: new bearings in history and the social sciences. Routledge, New York, pp 133–150

Bodenhamer DJ (2008) Creating a landscape of memory. Int J Humanit Arts Comput 1(2):97–110

Bodenhamer DJ, Corrigan J, Harris TM (eds) (2010) The spatial humanities: GIS and the future of humanities scholarship. Indiana University Press, Bloomington

Carter P (1987) The road to botany bay. Faber and Faber, London

[16]For another view of how the use of Web 2.0 tools will reshape traditional notions of authority, see Johnson (2011). Also see the various essays in Scharl and Tochtermann (2007).

de Certeau M (1984) The practice of everyday life. University of California Press, Berkeley
Cresswell T (2004) Place: a short introduction. Wiley-Blackwell, Malden
Daniels S, DeLyser D, Entrikin JN, Richardson D (eds) (2011) Envisioning landscapes, making worlds. Routledge, New York
Dear M, Ketchum J, Luria S, Richardson D (eds) (2011) Geohumanities: art, history, text at the edge of place. Routledge, London
Doorn P (2005) A Spatial turn in history. GIM international 19(4). http://www.gim-international.com/issues/articles/id453-A_Spatial_Turn_in_History.html. Accessed 28 Oct 2011
Gaddis JL (2002) The landscape of history. How historians map the past. Oxford University Press, Oxford
Giddens A (1984) The constitution of society: outline of the theory of structuration. Oxford University Press, Oxford
Harris TM, Rouse LJ, Bergeron S (2010) The geospatial Semantic Web, Pareto GIS, and the humanities. In: Bodenhamer DJ, Corrigan J, Harris TM (eds) The spatial humanities: GIS and the future of humanities scholarship. Indiana University Press, Bloomington, pp 124–142
Haulttunen K (2006) Groundwork: American studies in place. Am Q 58(1):1–15
Johnson I (2011) Spatiality and the social web: resituating authoritative content. In: Dear M, Ketchum J, Luria S, Richardson D (eds) Geohumanities: art, history, text at the edge of place. Routledge, London, pp 267–276
Kern S (1983) The culture of time and space, 1880–1918. Harvard University Press, Cambridge
Knowles AK (ed) (2008a) Placing history. How maps, spatial data, and GIS are changing historical scholarship. ESRI, Redlands
Knowles AK (2008b) What could Lee see at Gettysburg? In: Knowles AK (ed) Placing history. How maps, spatial data, and GIS are changing historical scholarship. ESRI, Redlands, pp 236–260
Massey D (1999) Space-time, 'Science' and the relationship between physical geography and human geography. Trans Br Geogr Soc New Ser 24:261–276
Massey D (2005) For space. SAGE, London
Pearson M, Shanks M (2001) Theatre/archaeology: disciplinary dialogues. Routledge, New York
Peuquet D (2002) Representations of space and time. Guildford Press, New York
Priestnall G, Jarvis C, Burton A, Smith MJ, Mount NJ (2012) Virtual geographic environments. In: Unwin DJ, Foote KE, Tate NJ, DiBase D (eds) Teaching geographic information science and technology in higher education. Wiley-Blackwell, Oxford, pp 257–288
Scharl A, Tochtermann K (2007) Geospatial web: how geobrowsers, social software, and web 2.0 are shaping the network society. Springer, London
Schuurman N (2004) GIS: a short introduction. Blackwell, Oxford
Sewell WHJ (2005) Logics of history: social theory and social transformation. University of Chicago Press, Chicago
Sheppard E (2005) Knowledge production through critical GIS: genealogy and prospects. Cartographica 40(4):5–21
Tuan YF (1974) Topophilia: a study of environmental perception, attitudes and values. Columbia University Press, New York

Chapter 2
Immobile History: An Interview with Emmanuel Le Roy Ladurie

Alexander von Lünen and Emmanuel Le Roy Ladurie

Emmanuel Bernard Le Roy Ladurie (b. 1929) is most known for his work in early modern history of the peasantry of southern France. His doctoral thesis *The Peasants of Languedoc* (published in French in 1966 and in English in 1974)[1] was supervised by Fernand Braudel (1902–1985). At that time, Braudel was the informal chief of the second generation of the *Annales* school which was dominated by the quantitative study of regions.[2] Le Roy Ladurie is often called the "standard-bearer" of the third generation of the *Annales* school, which was named after the journal *Annales d'histoire économique et sociale*. Founded by French historians Marc Bloch (1886–1944) and Lucien Febvre (1878–1956) in 1928, the school became famous for its then radically new approach to history by combining geography, history and social studies.

While the *Annales* school is still carrying on, its influence is less prominent in historiography than it used to be, a demise partly self-inflicted by its success. What once appeared as revolutionary—to combine fields such as geography, social and cultural sciences, or anthropology—is a rather common phenomenon in contemporary academic history.

Emannuel Le Roy Ladurie taught at the *Lycée de Montpellier* from 1955 to 1957 and at the University of Montpellier from 1960 to 1963, before becoming director of studies at the *Centre de Recherches Historiques* (CRH) of the *Ecole des Hautes Etudes* in Paris in 1963. In 1970 he moved to the Sorbonne, and he became professor of demography and the social sciences at the University of Paris in 1971. After Fernand Braudel retired in 1972, Le Roy Ladurie attained his chair of history of modern civilisation at the *Collége de France* in 1973 and retired in 1999.

---

[1] Cf. Le Roy Ladurie (1974).

[2] See Burke (1990) for more information about the *Annales*.

A. von Lünen (✉)
Geography Department, University of Portsmouth, Portsmouth, UK

E. Le Roy Ladurie
Collège de France, Paris, France

A. von Lünen and C. Travis (eds.), *History and GIS: Epistemologies, Considerations and Reflections*, DOI 10.1007/978-94-007-5009-8_2,

His early work was by and large centred on quantitative studies and regional studies, and he was one of the first to use computers in historical research. In his book *The Territory of the Historian*,[3] he proclaimed that "tomorrow's historian will have to be able to programme a computer in order to survive" (Le Roy Ladurie 1979, 6). Between 1987 and 1994, he was also a director (*administrateur général*) at the National Library of France (*Bibliothèque nationale de France*), where he was responsible for the digitisation of the library's catalogue, among other things. When at the CRH, Le Roy Ladurie was involved (together with François Furet (1927–1977), also from the CRH) in a Franco-American project to digitise a three-volume compendium, the *General Statistics of France* (French: *Statistique générale de la France*), which includes census reports of France (1851–1921), data on education from 1800 to 1925, industrial surveys between 1861 and 1890, and more. Working in the United States at the Inter-University Consortium for Political and Social Research (ICPSR), located in Ann Arbor, Michigan, the team digitised the bulk of the statistical data. However, due to internal problems (which will be referred to in the interview), the data was not available until recently.[4] In 1975, Le Roy Ladurie wrote what was to become a popular and best-selling book *Montaillou*, a micro-historical study of the town's medieval history.[5]

Alexander von Lünen (AvL) spoke with Emannuel Le Roy Ladurie (ELRL) on 21 October 2011 in Paris, France, about computers and geography in history.[6]

AvL: Monsieur Le Roy Ladurie, you were a pioneer in computers and geography in history …

ELRL: No, in geography and quantitative history and helped by others with computers. I worked with computers, but I was more an intellectual director and not really a practitioner of computers. I worked mainly in Paris on French History from the fifteenth century to the Revolution to the 1870s. I used military service registers, tithes, climate records and general statistics of France as sources. It doesn't mean that these statistics were everything, but they were our main source.

AvL: So, you said you didn't do the computations yourself (or the programming). Did you have a team, or partners who cooperated with you?

ELRL: At the beginning, we used a private firm to do those things. After that … let me see …

He reaches for a book, browsing; it's Anthropologie du conscrit Français from 1972.[7]

---

[3]Published in French in 1973, in English in 1979 (Le Roy Ladurie 1979).

[4]Another issue was in the way the data was formatted, which contributed to making the situation worse. The data has recently been transformed into spreadsheet files and is available from http://acrh.revues.org/index2890.html (last access: 14 November 2011, page is in French) for download.

[5]English version: Le Roy Ladurie (1980b).

[6]This is an edited version of the original transcript. Interview held in English. See the entry for Le Roy Ladurie in Cannon et al. (1988) and Hughes-Warrington (2000) for the biographical details.

[7]See Aron et al. (1972).

ELRL: ... we had a very good laboratory. Braudel installed—now it's disappeared—a very good laboratory of cartography, which was directed by Jacques Bertin. At the beginning it was small and a bit underequipped, no computers at first.

ELRL asks AvL to come over to his side of the table, and shows and explains to me in detail the maps that are contained in the book. As we go through the book, I see a great amount of thematic maps on historic France, chiefly between 1815 and 1830 during the "restauration", with an impressive range of categories and variables. These include volunteers for the army, deserters, people who bought themselves out of military service, mining, industry, agricultural workers, average height of the male population, occupations, education, etc., for all French Departments.

ELRL: The idea of the book was a comparison between "developed vs. less-developed" regions, i.e. the level of economic and social development in the various regions.

AvL: How did you produce that book? I mean, were these maps done by hand?

ELRL: No, no, by computer.

AvL: Who produced them? A private company?

ELRL: No, that was done by the cartography lab we had.

AvL: And how did the cooperation work? Did you tell them what to do, or did you just give them the tables and they produced the maps from them?

ELRL: Yes. Bertin was an intelligent man, I gave him the data and he knew what to do. It is obviously an advantage to have 90 rather small departments in France, which allows relatively subtle geographies. When we had only the big provinces, like the Provence, you could still make sense of the data, but less so ... So maybe the division into departments shouldn't have been done by the French Revolution, but for the historians it is very handy.

AvL: And how did you produce the tables? Did you use computers for that?

ELRL: There was a group of young people who did that.

AvL: At your university?

ELRL: At the *Ecole des Hautes Etudes*, yes.

AvL: So, basically, you took the numbers from the archives...

ELRL: I made the samples, and when we had enough samples, we gave it to the right person. ... I think the other big work which has not given great results, but is still important, is *Statistique générale de la France*. It was an idea I had with Jacques Dupacquier and François Furet, and I was in Ann Arbor, Michigan, USA, at the time ... and in Ann Arbor they had this Consortium; and I went over and gave them the idea of putting *Statistique générale de la France* into the computer. As a centralized country France had really great statistical data, even since 1820, and of course since the 1850s. So, the Ann Arbor people found the money, American money—at the time there was no existing competition for such a topic. And we had young French students, people who were paid for that, who photocopied the big volumes of the *Statistique générale de la France* for the Americans and we started, as far as I can remember, with two research agendas:

one was historical demography; the other one was history of education. Because, at the time, one of our main interest was in education. Furet was not so much interested in my approach to demographic and social problems, so with Ozouf and other scholars, he published two volumes on primary education in France.

But then the catastrophe of 1968 happened; everything was registered and digitized, but it remained in the archives of the consortium of Ann Arbor, and partly in France ... and finally some stupid young man delivered a lecture where he said that Furet had used the money of the state for no serious purpose ... so I had to recollect the archive and return to France; and the archives—the digitized archives—of the *Statistique générale de la France* have been available in America for a long time now, and also in the Centre of Historical Research in the *Ecole des Hautes Etudes en Sciences Sociales*. However, some young men are starting afresh with this material. I am very happy with that.

And the problem for me was that Braudel was not really interested with those quantitative archival developments. I was not in disgrace, but he wasn't happy; actually, I remained on good terms with him, but I got a professorship at the Sorbonne, an institution he did not like too much, and when I changed to the *Collége de France*—replacing him in his own professorship—I was a bit separated from the French universities as such. So I ... I did not abandon historical demography, but rather focused on history of climate, which I have continued since, together with other subjects. Furthermore, the young people who were doing the computer-related work for me received tenure. The result was that they went to do qualitative history. And so to some extent I had to switch topics.

AvL: You mentioned "the Catastrophe of 1968"?

ELRL: I think [the student rebellion of] 1968 was a disaster, from my point of view, because it ... the young people are no longer interested in all those statistics. We were not econometricians, but we dealt resolutely with figures.

AvL: So the breakup of your team made you break away from most of your quantitative studies?

ELRL: The problem was that I was no longer the real director of the laboratory called *Centre de Recherches Historiques* at the *Ecole des Hautes Etudes*. Furet became the new real leader of it and he was no longer interested in quantitative history at all, after his interesting book on literacy. He simply used with—great talent—the old figures of Maggiolo and then left the field, quiet legitimately. I was a bit isolated at the *Collége de France* but not unhappy at all. So I explored different fields: demographical history, economical social and general French history.

AvL: Well, the main reason I was asking that question was because in *The Territory of the Historian* you said "Historians should become programmers to survive". Yet you never programmed the computers yourself?

ELRL: Well, I mean ... I shouldn't have said that. Although I think that the final aim of history, if it is not a purely cultural history, is to be quantitative wherever it is possible; but that is the final aim, the validation, not all the research has to be quantitative... So I should not have said that.

AvL: All history should be quantitative in the end? That's a remarkable statement, given that you're considered one of the champions of micro history since the 1970s, which was started from the premise that one can do social history without resorting to statistics.

ELRL: Well, I came to that by chance, basically. Yes, I had this surprise of *Montaillou*; that is: discovering through a local scholar—Jean Duvernoy—the existence of a register of the inquisition conducted by Jacques Fournier, bishop of Pamier in the Pyrenees in the fourteenth century; Fournier became pope by the name of Benedict XII later. And I was struck by the extraordinary quality of this collection, a meticulous record of the town's gossip: a certain woman said A, this man has said B. I discovered that under the weight of indirect statements of the "culprits", there were real dialogues between ordinary people.

Duvernoy, who knew Latin very well, had published three volumes of this inquisition register in Latin, plus a small study in French. And he had understood that it was important to study dialogues ... as an anthropological study ... and he published some of the dialogues in a small book.[8] But I realized that Duvernoy had not managed to immerse into the life of a particular village. So, at first a friend of mine and I made a film of this story for the French TV and that film had some success. And finally I decided to write a book: *Montaillou*.

I worked hard on the original Latin text. I always quoted the source, but of course Duvernoy was very surprised—just like me—that without any intent, I have inadvertently written a best-seller from his transcripts. But we remained on good terms, since thanks to me he could publish —well after my *Montaillou*—a complete translation in French of those three Latin volumes, which endowed other scholars with a remarkable edition and references.[9] The sources contain extraordinary narratives of love affairs, social life etc., but what is interesting about this book is that it is a register. This register is equally admirable as *Tristan and Isolde* as a great and unrivaled magnificent piece of literary work—especially in the Latin text—the content is absolutely incredible. My publisher, Gallimard, published it and then an unexpected success came, just by "from mouth to ear", not much public relations work was done. And on many of the beaches of Southern France you could see people reading the book. I think in France it sold 200,000 copies, and in England they still sell 1,000 copies every year; unfortunately, the translation is good, but a bit defaced.

[8]Cf. Duvernoy (1966).

[9]Cf. Duvernoy (1977–1978).

And, of course, when you write a best-seller and you are an academic ... it takes 20 years to recover from this occasion. But the Americans have been very ... you know I am pro-American, actually, but they made a film from it without my permission. It is a good film as such, but it is bad because they only speak up against the Catholic Church in it. *Montaillou* was a study of the province. And the systematic study of villages was an anthropological specialty, which I worked on when I was in America.

AvL: So the main reason you ventured into micro history was because you were intrigued by that particular narrative of Fournier?

ELRL: Yes.

AvL: It reflects a trend, though, in historiography. In the 1950s and 1960s, a lot of historians—if not the majority—worked with quantitative methods, and then in the 1970s people turned away from that ....

ELRL: Well, the idea was: one has to be quantitative, because many of us came from the French Communist Party when we were 20. Furet, me, and others wanted to have a materialistic approach, but to make it more quantitative (often with a bit primitive methods). So I started from tax registers and cadastres, because of the data that was available; and also there was a cooperation with geography at the time because of my great friend Dugrand, who made a very good thesis on the geography of vineyards with remarkable maps of the Bourgoisie and its possession of the vineyards, and that was a great inspiration for me.[10] And then I started to work on cadastres and demographic history of the sixteenth century; not especially with Parish registers, which did not exist for the time, but with numbers of people who paid taxes, shepherds and their families, for example. And a discovery of mine was the growth of population in the sixteenth century after all the deaths in the 1550s and 1560s ... and the thing I discovered that was entirely unknown to French historians was the demographic growth between 1460 to the beginning of the religious wars. Then I started from the Marxist view of concentration of the land in the hand of the capitalists, something that was also a popular topic in English history at the time, of course.

I started from this idea, and then I discovered that no big changes took place. Basically the population grew, and the farmland was divided into smaller parcels, in the style of older French agriculture, and when the population became stable in the seventeenth and eighteenth century, there was some slight accumulation of wealth. I think that was my second discovery and that is why from Marx I went to Malthus. That is the idea of high population that meets limits of growth. Of course, much of it is too simple, it's only food that imposed the major limit; there are also epidemics and war, of course, which is a great limitation; and then, I think, I worked on tithes, which gave me a view of history of agricultural production

[10]Cf. Dugrand (1963).

and collective rents; and there was some interesting controversy, because tithes are difficult to quantify for fluctuations within 1 year, of course; and then there are climate changes: 1709, hard winter, agricultural production plunges and also the next 10 or 20 years; in regard to the long term changes, however, it is not so evident. And it's very interesting to see that during the eighteenth century, the amount collected by tithe does not increase; but actually, if you look at France the population from 1715 to 1789 was growing from 20 to 28 million, and the famines had been even more important in the second half of the seventeenth century, I mean the famine of 1661, famine of 1696, famine of 1709, but after that famines more or less disappear. So it's impossible that you could—of course the mills etc. became more efficient—have an important demographic growth and no serious increase in agricultural production. So you see that one has to be careful with a conclusion from the data.

AvL: So you dealt a lot with social and economic structures in your thesis and other works. Your work—and that of Braudel—was heavily concerned with "structure" and longer temporal scopes. In how far does *Montaillou* fit in there?

ELRL: Well, Braudel was quite influenced by the German population history school. It played a vital part for his masterpiece on the Mediterranean. One major reason for that was his time as a prisoner of war in Germany during World War II. The Germans allowed French prisoners to use the books of the closest academic library, and Braudel was imprisoned near Hamburg, so he used the library of the university there. After that period, he worked on capitalism, first on a global and national level, then going down into the French regions. This approach, to tackle both the macro- and the micro-level has always been with the *Annales* school, actually. Marc Bloch, for instance, worked on the agrarian history of France, from macro-economic models as they were used in England and Germany. He started with an economic model and applied that to French rural history. Lucien Febvre, on the other hand, did "history from the grassroots", i.e. starting from small communities and then widening the geographical scope.

AvL: By "structure", do you refer to social structure or to geographical structure?

ELRL: Yes, that's where the different levels of structure and scope meet. I was interested in linguistic regions to begin with. You know, all the different dialects and languages in France, such as Occitan or Basque. I looked at the literature of southern France, at the theater plays in the southern French dialect between the seventeenth and nineteenth century. The Occitan language, when it is used in troubadour, sounds like a foreign language to a French person. From that I began to look at anthropology, especially the work of Claude Lévi-Strauss on myths and popular tales; the idea of transformations. And also his work on demography and family structures were an inspiration for me. So I made a book on all the non-French speaking areas in France, like Alsace etc. After that I wanted to "get out of France"—thematically. I came across the Swiss-German family of the Platters, to whom I have dedicated three books then, and my new publisher said: That interests nobody. But actually, I sold a bit, a little bit ... and the Platter family books I designed as family history ... but also as a geographical study, because of their travels. One of them travels as a beggar—very rough conditions—in Germany; and the two sons have made ... you can say "geographical history" ... that is: the oldest, which is the more intelligent one, Felix Platter, has made a journey as a student in Montpellier, and he described the Languedoc, Provence, Occitan, wonderfully with great affection and idealism and then he traveled through the rest of France—he made a Tour de France, actually—from the South-West through the Normandy, Paris and coming back through Gascoigne. So I made three volumes of that, and I think, it is also a geographical approach, because for each city, for each place that the historian may think is interesting I took the books and articles—not all of course—that have been published on those places, and I tried to enrich the narrative of Platter. And that is part of a long Southern experience, with *Montaillou*, which was a great success; and *Carnival de Romans*,[11] which was a normal success. And each time I was plundered by bad film-makers.

AvL: Very interesting. It looks like geography always played a vital part in your studies, both the quantitative and the micro-historical studies. How did you get interested in geography in history? I know it is quite common in France ...

ELRL: You mean geography and history in conjunction? Well, I think in Germany it is history and philosophy that are bonded, and in France for a long time now it is history and geography. I received training a little bit in geography as a student; geography is a highly esteemed subject in high-school in France, with very good teachers. But your question was how this approach developed in my career as historian?

AvL: Yes. You actually studied a bit of geography?

ELRL: Yes. I was also interested in climate history, which was to some extent geographical; a comparative history, which was started by Marc Bloch and some German historian—of which I can't recall the name. In the work of Braudel the Mediterranean was obviously important, and that also influenced me a great deal.

AvL: I know that in France history and geography are very close. Is there a particular reason for that, historically?

ELRL: I think ... the original reasons I don't know very well, but I consider that in terms of intellectual history, it is the geographical focus at the Sorbonne ... I think the French geographical school at the beginning of the nineteenth century was very well developed with Vidal de la Blache (geography of France); and the starting point of the *Ecolé de Sedan* was

[11] Cf. Le Roy Ladurie (1980a).

with Lucien Febvre and Marc Bloch. Lucien Febvre wrote a big thesis on the Franche Comte in the time of Phillip the Second. You know Franche Comte, the region between Switzerland and Germany. And it was a very bright book on the grassroots of history; it transported the methods of the French geographical school to history. Of course there's a tendency, there are trends, but it is a real geographical description of the province in the style of a geographical survey; and it is a very innovative book, radical almost. He published his thesis in 1912 and it generated the big book by Braudel *Mediteranée*... .[12] Now, essentially, after the war we all started geographical history of regions; which, of course, was a bit provincial, but with a lot of good books.

AvL: Yes, I noticed. I mean, I am not an expert on French history, but I noticed that a lot of books are on regional history, whereas in Germany that is not a very prestigious topic.

ELRL: Yes, it has its shortcomings, of course. It is provinciality, but the studies were quite deep.

AvL: Was that maybe because France was so centralized? I mean, more of a centralized state than, say, Germany, which was first regional and then was unified into a nation state rather late in 1871, so people looked more at the process of nation-building.

ELRL: It is not the only reason. Because it is a centralized country, it has excellent regional and departmental archives. You know, because of the division into 90 departments in the wake of the French Revolution, and the duty of every department to report its economic and social data to the central government. For this reason, you can make a thesis about agriculture in the nineteenth century in the department of Calvados, for example, with relative ease. But I agree with you, when you have for instance the history of World War II or the Nazi regime, this is so important that you can neglect the regional history of Bavaria etc. That is my impression; that goes in your direction.

AvL: How do you see the relationship between History and Geography in general?

ELRL: There are of course similarities, but I can't say much about the theoretical underpinnings. Geography has always informed me a great deal. I mentioned my demographic and micro-historical studies already, which had a strong geographical impetus. But there is also a concept I developed later: that of "Immobile History", an extension of Braudel's *Longue Durée*. I looked at the fourteenth century when France was not so populated. There were 20 million French at the time. And I observed that the South resists major changes through time, that is: radical changes. Population increases prevail. After all the killing in the seventeenth and especially

[12] See Braudel (1966) for the French 2nd edition, or Braudel (1972) and Braudel (1973) for the English translations. This book by Braudel is a milestone in spatial history (among other things).

eighteenth century, the population grows; and then there are geographical expansions, city expansions, etc.; but the basic thing was: the region has not changed so much in many respects in that time. Ok, there are things like Catholicism stalling—Protestantism has never been very big, anyway ... strangely enough, Protestantism in France is like rugby, it's a Northern specialty, but it is not seen in the South. So the religion is still Catholic, languages are still the same: French—some kind of French—different from the language in the North, and all the languages like Celtic, Occitan, etc. Agriculture has been improved, but is relatively constant: ox in the North, ox in the South, two field rotations in the South, three field rotations in the North—subsistence agriculture. Of course, farmers became more efficient, but the proportions in agriculture remain the same. Urbanization doesn't change too much, either. Most of the 30,000 or so villages are there since the eleventh century. Setting occasional dents in the curve aside, we still have around 20 million people in 1705—so there was no variation over a long time by and large ... and that is what I call "immobile history". Of course, it neglects trends like growing cities; they had a greater part of the population, maybe 60% or a little bit more. But the general social and economic trends prevail.

AvL: Interesting. You really had a broad range of interests and approaches. We don't have that anymore so much. People are very specialized and ignore a lot of things these days. Saying: That's out of my boundaries.

ELRL: *Non*? Except when you're over 60. I have been called a polymath once, but that's an exaggeration. I was just interested in a broad range of topics and approaches. I would have loved to do more with computers myself, but when I started my work we didn't even have programmable computers, just calculating machines. Nowadays all the historians have computers and laptops, but they don't use them in their research really.

AvL: I couldn't have said it better myself. *Merci beaucoup.*

References

Aron JP, Dumont P, Le Roy Ladurie E (1972) Anthropologie du conscrit français d'après les comp. Mouton, Paris

Braudel F (1966) La Méditerranée et le monde méditerranéen a l'époque de Philippe II, 2nd edn. Colin, Paris

Braudel F (1972) The Mediterranean and the Mediterranean world in the age of Philip II, vol 1. Collins, London

Braudel F (1973) The Mediterranean and the Mediterranean world in the age of Philip II, vol 2. Collins, London

Burke P (1990) The French historical revolution: the annales school, 1929–1989. Polity, Oxford

Cannon J, Davies RHC, Doyle W, Greene JP (eds) (1988) The Blackwell dictionary of historians. Blackwell, Oxford

Dugrand R (1963) Villes et campagnes en Bas-Languedoc: le réseau urbain du Bas-Languedoc méditerranéenne. Presses Universitaires de France, Paris

Duvernoy J (1966) Inquisition à Pamiers. Interrogatoires de Jacques Fournier, 1318–1325, choisis, traduits du texte latin et présentés. Privat, Toulouse
Duvernoy J (ed) (1977–1978) Le Registre d'inquisition de Jacques Fournier, évêque de Pamiers: 1318–1325. Civilisations et sociétés 43, La Haye/Mouton, Paris/New York, 3 vols. Préface de Emmanuel Le Roy Ladurie
Hughes-Warrington M (2000) Fifty key thinkers on history. Routledge, London/New York
Le Roy Ladurie E (1974) The peasants of languedoc. University of Illinois Press, Urbana/Chicago/London
Le Roy Ladurie E (1979) The territory of the historian. Harvester, Hassocks
Le Roy Ladurie E (1980a) Carnival: a people's uprising at Romans 1579–1580. Scolar, London
Le Roy Ladurie E (1980b) Montaillou: Cathars and Catholics in a French village, 1294–1324. Penguin, Hammondsworth

Chapter 3
The Dawn of a Golden Age? Historical GIS and the History of Choropleth Mapping in the Netherlands

Onno Boonstra

3.1 Introduction: A Perfect Storm

The nineteenth century is generally considered to be the period during which the many facets of modernization started to take shape in Europe and North America, only to expand throughout modern society within the last half of the twentieth century. Of course, this does not mean that all social, cultural, economic, and technical innovations during the nineteenth century developed in the same pace or in the same direction. There have been developments that were basic building blocks for the modernization process but ceased to be of importance in later years; there have been other developments that were also basic building blocks for the modernization process but had their finest hour in the nineteenth century.

Michael Friendly, who studies the history of data visualization, claims that the latter holds true for thematic cartography. For sure, we cannot deny living in a golden age of cartography at this very moment because indisputably the importance of mapping in everyday life has never been of higher value than it is today. Hardly anyone will have trouble to find their way through the "real world," be it with the help of GPS navigation systems or Google Maps. But this should not construe that the halcyon days of thematic cartography exist today as well. On the contrary, according to Friendly, the golden age of thematic cartography was more than a century ago, between 1850 and 1900. This is because during that period of time, developments in statistics, data collection, and technology converged to produce "a perfect storm" which allowed the emergence of statistical graphics and thematic cartography (Friendly 2008).

To some extent, Friendly is right: During the second half of the nineteenth century, statistics had established itself as an important social science (Potter 1986).

O. Boonstra (✉)
History Department, Radboud University Nijmegen, Nijmegen, The Netherlands
e-mail: o.boonstra@let.ru.nl

A. von Lünen and C. Travis (eds.), *History and GIS: Epistemologies, Considerations and Reflections*, DOI 10.1007/978-94-007-5009-8_3,

Alongside, data collection had taken shape through surveys and censuses, and lithograph printing technology had advanced in such a way that publishers were able to issue multicolored maps at a moderate price (Woodward 1975). However, these developments did not start until around 1850; the increase and improvement of statistics, data collection, and printing techniques had been set in motion already decades before. As a consequence, there are examples of thematic cartography which were produced a very long time before the dawn of the golden age of cartography the most famous being Charles Dupin's (1784–1873) map from 1826 on school enrollment in France.[1] Obviously, statistics, data, and technology were not the only conditions that needed to be met in order to enter a golden age of thematic cartography.

Therefore, the history of thematic cartography and the prerequisites that were needed to advance toward its golden age may serve as an example for the development of historical GIS in our time.

3.2 The Invention of Choropleth Mapping, 1826

In the history of cartography, the "invention" of thematic mapping is a major milestone (Friendly 2011). It was a real discovery to consider a map not merely as a set of points, lines, and polygons in order to visualize geographical features but as a carrier of quantitative data which did not give information about the geographical features represented on the map but instead provided information about regionally differentiated non-geographical features.

Apart from a few very early attempts in the seventeenth and eighteenth centuries, a conscious use of thematic cartography only gained weight when statistics entered the scientific domain in Europe. Subsequently, quantitative data started to be collected, arranged, statistically summarized, and analyzed in a systematic way (Koeman 1983). At the end of the eighteenth and the beginning of the nineteenth century, the number of reports and books based on statistical data started to grow tremendously in both Western Europe and the United States. Predominantly, traditional approaches such as tabularization were used to present statistical results. Tabularization was the most commonly used statistical tool at that time to present data. It had in fact been in use for decades, for instance, in the presentation of life expectancy statistics used to calculate yearly deposits for life insurances (Klep and Stamhuis 2002).

But, apart from tables, very gradually, a new way of summarizing and presenting statistical data began to emerge at the end of the eighteenth and the beginning of the nineteenth century: visualization (Palsky 2004). Timelines, plots, and bar charts gradually made their way into statistics (Beniger and Robyn 1978). To our

[1]See the interview with Emmanuel Le Roy Ladurie in this book for a bit of information on the French geographical school.

contemporary eyes, it may seem to be a small step to move from one visualization technique to another and to use, for instance, a thematic map as a visualization tool for quantitative data. However, it took more than 25 years before this step was taken, and thematic mapping was "invented" (Klep and Stamhuis 2002). It is recognized that the first choropleth thematic map in the world was made by Charles Dupin in 1826, the *Carte figurative de l'instruction populaire de la France* (Dupin 1826). Dupin published the map as a result of a series of university lectures he had delivered at the *Conservatoire des Arts et Métiers in Paris* (Palsky 1996). The *Carte figurative*, showing marked departmental differences in school enrollment in France, was accompanied by a number of tables and remarks that pointed toward the large educational and therefore social and economical progress of the north of France when compared to Southern France.

3.3 Choropleth Mapping in the Netherlands, 1827–1865

Dupin was a trained engineer who had mastered skills in both statistics and cartography. It made him the ideal person to synergize these skills into the invention of choropleth mapping. But, being an engineer, Dupin was an outsider. By profession, he was neither a statistician nor a cartographer. As a result, he was not very successful in stirring enthusiasm in others, be it statisticians or cartographers, to follow his example. Only Guerry followed Dupin's footsteps in 1829 and 1833 by drawing sets of choropleth maps, showing regional differences in education, crime, illegitimate births, suicide, and donations to the poor (Friendly 2007). The most striking imitation of Dupin's work, however, came from the Netherlands, where the second choropleth map in the world was published: the *Carte figurative de l'instruction populaire des Pays-Bas*, published in 1827 (see Fig. 3.1).[2]

The data which is visualized on the *Carte figurative de l'instruction populaire des Pays-Bas* is derived from a large survey into regional variations in educational enrollment in the Netherlands that had been ordered by the Dutch *Ministerie van Binnenlandse Zaken in* 1825. In each and every of the 3,000 municipalities in the country, a count had been made of all children who went to school, be it to a primary school or to a nursery school or be it in wintertime or summertime.[3] The results of the survey were published in an elaborate report in 1826 (Anonymous 1842). As had become customary during those years, the data had been presented and summarized with a very extensive set of tables. The report was published in Dutch, but a translation of the results of the survey into French was published one year later as a service to the large minority of Francophones in the country (Anonymous 1827). Alphonse Quételet summarized the French report for an analysis of his own

[2]Somerhausen (1827). More about this map in Boonstra (2009a).

[3]From 1815 to 1830, the Kingdom of the United Netherlands contained the area of nowadays' the Netherlands, Belgium, and Luxemburg.

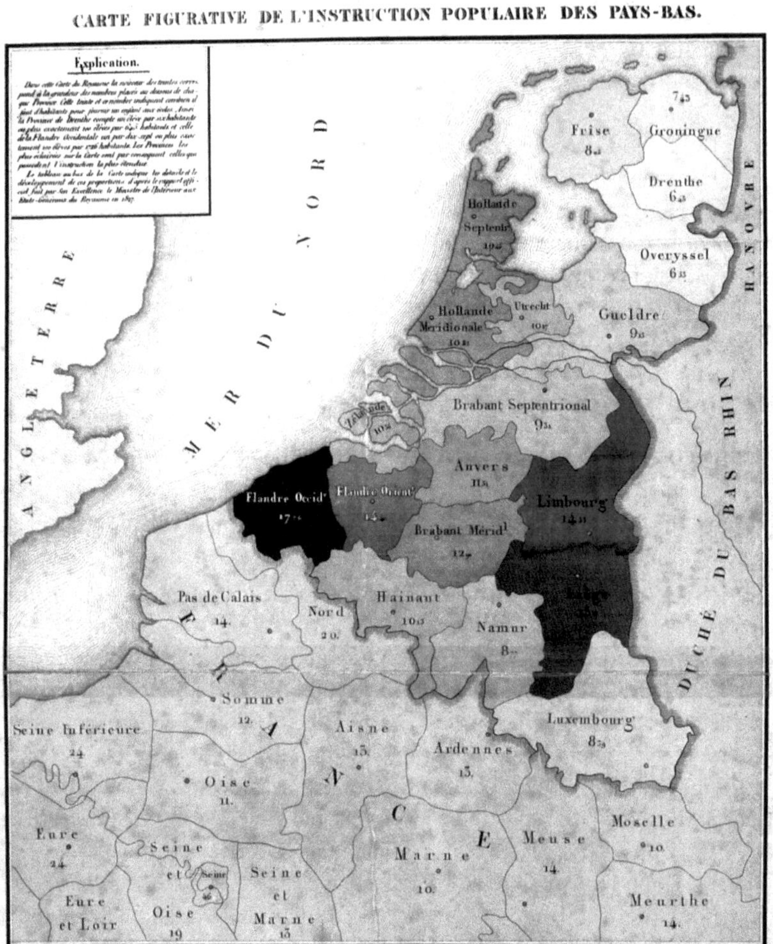

TABLEAU de l'état de l'Instruction dans les différentes provinces du Royaume des Pays-Bas, d'après le rapport officiel fait aux États Généraux, en 1827.

Noms des Provinces d'après l'ordre suivi dans la loi fondamentale. (Celles suivies d'un S. sont septentrionales, d'un M. méridionales, la petite étoile indique les prov. walonnes)	Population.	Nombre d'élèves sur 1000 habitans des villes et communes d'une population			Nombre des élèves jouissant de l'instruction dans les établissemens publics et autres, les petites écoles exceptées.	A. Proportion de ces élèves, sur 1000 de la population	Nombre des enfans qui fréquentent les petites écoles.	Total des élèves, admis dans les établissements d'instruct.on en général.	B. Proportion de ces élèves s. 1000 de la population.	Population jouissant d'instruction ailleurs que chez elle.	Population privée de toute instruction.	Rang des provinces, d'après la proportion dans la colonne A.	Rang des provinces, d'après la proportion dans la colonne B.	Étendue en hectares (bonniers) d'après le tableau officiel présenté aux É. G. le 20 déc. 1826.	Nombre des	
		au delà de 6,000 âmes.	entre 6000 et 1200.	de moins de 1200 âmes.											villes dans chaque province	communes et villages.
1. Brabant septentrional . S.	527,526	77.80	100.51	111.75	55,159	107.41	2,819	37,978	116.02	926	—	8	8	484,896	10	174
2. Brabant méridional . . M.	492,756	56.38	74.88	80.68	38,170	77.46	3,571	45,541	88.76	1,252	16,307	15	15	507,733	8	524
3. Limbourg . . . . M.	524,368	65.81	66.56	66.10	22,288	68.71	1,466	23,754	73.25	—	64,008	17	18	455,316	9	538
4. Gueldre . . . . S.	254,266	91.60	98.02	111.85	51,124	109.49	2,031	33,155	116.63	—	—	6	7	517,098	13	111
5. Liége . . . . *M.	337,556	58.90	68.43	64.19	22,333	66.16	1,000	23,333	69.12	4,822	28,919	18	19	282,593	7	325
6. Flandre orientale . . . M.	689,158	84.38	59.43	61.57	47,849	69.43	8,023	55,872	81.07	4,283	48,651	16	17	298,370	11	308
7. Flandre occidentale . . M.	671,034	94.69	72.23	39.74	38,858	57.90	18,264	57,122	85.12	2,808	6,359	19	16	517,422	15	233
8. Hainaut . . . . *M.	546,245	92.53	100.55	93.20	53,915	98.70	6,522	60,437	110.64	1,177	50,689	9	12	577,390	21	429
9. Hollande septentrionale S.	405,929	88.00	128.52	98.93	38,898	95.82	9,150	48,048	118.86	271	487	12	6	229,200	11	155
10. Hollande méridionale . S.	440,662	87.42	94.65	89.73	40,698	92.35	9,475	50,173	113.40	6,495	6,502	13	10	277,830	13	257
11. Zélande . . . . S.	132,321	86.95	91.12	100.17	12,772	96.52	1,433	14,205	107.35	973	1,988	11	13	158,036	9	116
12. Namur. . . . . *M.	199,482	101.34	121.26	105.75	21,704	108.56	1,274	22,978	115.18	2,353	24,056	7	9	345,610	5	337
13. Anvers . . . . M.	319,285	66.67	81.00	96.06	27,719	86.81	3,682	31,401	98.35	1,346	216	14	14	282,293	4	150
14. Utrecht . . . . S.	122,395	90.40	90.17	103.12	11,921	97.39	1,745	13,666	111.65	5,034	3,482	10	11	127,617	6	92
15. Frise. . . . . S.	200,654	109.80	117.18	128.64	24,922	124.20	2,011	26,933	134.22	7,256	—	4	4	260,732	11	146
16. Overyssel. . . . . S.	157,158	157.91	170.70	152.58	24,071	153.16	1,801	25,872	164.62	—	200	2	1	329,961	5	60
17. Groningue . . . . S.	157,973	151.55	129.98	135.81	21,257	134.56	331	21,588	136.65	97	—	3	3	205,059	1	56
18. Drenthe . . . . S.	56,979	—	178.07	147.55	8,809	154.60	90	8,899	156.18	—	—	1	2	225,852	3	30
19. Luxembourg . . . . . . *M.	291,759	111.26	112.69	118.15	34,744	119.08	160	34,904	119.63	759	9,728	5	5	626,343	18	317
TOTAUX	6,157,286	—	—	—	557,211	—	76,648	633,859	—	40,052	241,592	—	—	6,107,351	180	3,938
Nombres moyens	—	91.28	101.75	99.29	—	90.49	—	—	102.94	—	—	—	—	—	—	—

NOTE. Voir, pour les Conclusions à faire sur le plus ou moins de lumières dans une partie du Royaume que dans l'autre, les judicieuses Observations de M. le Professeur *Quetelet*, dans ses Recherches sur la population, etc. Bruxelles, page 65 et suivantes.

Fig. 3.1 The *Carte figurative de l'instruction populaire des Pays-Bas*, 1827 (With permission from the University of Utrecht library)

to be included into a much more verbose statistical essay on social phenomena of that time (Quételet 1827). Among other things, Quételet looked into the ratio between the number of children who go to school in each province and the number of inhabitants in each province. The ratios showed that the provinces of Drenthe and Overyssel had the lowest ratios, meaning that in those provinces the number of school pupils was relatively high. One in every 6.5 inhabitants went to school in those provinces. Other provinces that performed almost equally well were Groningue and Frise. Bad performing provinces, showing high ratios, were Flandre Orientale, Limbourg, and Liège but above all Flandre Occidentale, which had the worst ratio of all provinces: one pupil for every 17.3 inhabitants. The result seems to point toward a marked division between the Northern and Southern Netherlands, but this is only partly so. The map, for instance, illustrates that the southern, French-speaking provinces of Luxembourg, Namur and Hainaut had better ratios than the northern provinces of Hollande Septentrionale, Hollande Méridionale, and Utrecht. Another rather striking outcome of Quételet's analyses is that the provincial results reflect only to some degree the generally accepted idea that school enrollment in the Netherlands in the nineteenth century had been seriously influenced by religious affiliation, causing predominantly Protestant provinces to show better ratios than Catholic provinces. And although the results quite convincingly make clear that in Protestant provinces like Drenthe and Groningue many children went to school, they also show that this was also the case for the religiously mixed province of Overijssel. And in Brabant Méridional, a province with an overwhelmingly Roman Catholic population, school enrollment in 1825 was higher than in the Protestant province of Hollande Septentrionale.

However, in the first half of the nineteenth century, most statistical inquiries did not meet the demands we set to them today. Not only were surveys used to pursue the political goals of the nation and the national elites, but also the quality of the data that were gathered sometimes was insufficient (Woolf 1989; Schwebber 2006; Randeraad 2002). This was true for the 1825 survey. For instance, the number of school pupils that have been counted in the province of Overijssel was higher than the total number of children in the 6–12-year-old age group.[4] Quételet himself questioned the very bad results for the province of Oost-Vlaanderen (Quételet 1827, 66). It also seems that the word "school" has not been employed uniformly. A differentiation had been made between primary schools and nursery schools (*petites écoles*), but the number of children attending nursery school was so very different for each province that one can argue that in some provinces, most of the nursery schools must have been counted as primary schools.

Although the results may be questioned, the *Carte figurative de l'instruction populaire des Pays-Bas* remains remarkable for three reasons. First of all, it is astonishing to see how much the Dutch map is a copy of Dupin's map of school

---

[4]This calculation has been made by dividing the number of children enrolled into school at 1 January 1826 by the number of children aged 10–16 years (and therefore representing the number of children aged 6–12 in 1826) according to the census of 1830.

enrollment in France, which had been published a year before. Not only the design of the map—the title, the tabular account underneath the map, the type face, and the way the map had been shaded—is similar but even the variable which is depicted on the map. In both cases, the maps show the number of pupils enrolled in primary education compared to the number of inhabitants. Even Dupin's data on French educational attainment have been added to the map of the Netherlands. It not only makes clear how much the cartographer was impressed by Dupin's pioneering work but also that he had grasped the tremendous possibilities of thematic mapping, in particular the possibilities that choropleth maps have in showing regional variations.

The second reason why the map is remarkable is that the *Carte figurative de l'instruction populaire des Pays-Bas*, although based on data provided by Alphonse Quételet, was not Quételet's own work. The Carte figurative was published by a man called Hartog Somerhausen. Somerhausen was neither a statistician nor a cartographer; he was a Jew from Bavaria who had immigrated to Brussels in 1817, where he had become a leader of the local Haskalah, the Jewish Enlightenment movement (Kasper-Holtkotte 2003, 271). He founded a grammar school for impoverished Jewish boys, joined the Dutch-speaking Enlightenment organization Concordia, worked as a translator and wrote—in his spare time—on the history of Brussels and Belgium.[5] Quételet was an innovative statistician, who looked beyond his own discipline, and not only introduced new statistical methods, techniques, and research themes but also new visualization techniques. But although Quételet knew about Dupin's work, he did not take the initiative to visualize his own calculations cartographically. It may be that he did not attempt to make a thematic map because he did not trust the survey's results. But it is more likely that he, despite possessing an open mind, did not conceive the possibilities of choropleth mapping.

This leads us to the third reason why Somerhausen's map is remarkable. Not only is it the second example of a choropleth thematic map in the world, but even more remarkable is that it took 30 years, until 1857, before another choropleth map was published in the Netherlands. It was in that year that a set of 16 maps was published by Jacob Kuyper in the German journal *Petermann's Mitteilungen* (Kuyper 1857). Fourteen of them were choropleth maps, showing differences between the eleven Dutch provinces in population density, religion, trade, education, and agricultural produce. Eight years later, in 1865, Gregorius Mees published a third choropleth map, the first one that was drawn and printed within the borders of contemporary Netherlands. It was a small inset showing the population density of the Dutch provinces according to the census of 1859.[6]

---

[5] Weijermans (2010) and Frederiks and van den Branden (1888–1891, 741)

[6] Mees (1865, 13). The map was called "De betrekkelijke dichtheid der Provinciaale Bevolking door schaduwen" (Engl: Relative population density of Dutch provinces shown with shadows). On the first thematic maps that were published in the Netherlands, see Koeman (1983, 240–254), Wallis (1989).

3.4 A Fourth Prerequisite for the Golden Age of Thematic Cartography

There is a striking similarity between the Dutch pioneers of thematic cartography Somerhausen, Kuyper, and Mees. All of them, just like Dupin, were outsiders to the disciplines of cartography and statistics. At the time when his first maps were published, Jacob Kuyper (1821–1908) was a tax collector with an amateur's interest in geography.[7] Gregorius Mees (1802–1883) was a lawyer and a historian. Already in 1840, when he was appointed professor in Dutch history at the Deventer Athenaeum, he pleaded not only to improve the quality of the research into historical sources and to reserve funds to annotate and publish them, but most of all he argued for the use of historical maps in order to summarize historical events or circumstances (Mees 1840). The Historical Atlas of the Netherlands he so much desired was eventually published 25 years later—by himself.

As Koeman writes about the *Carte figurative de l'instruction populaire des Pays-Bas*: "The map is a very solitary one. It would take until the 1860s before a second choropleth map was published in the Netherlands. In the meantime, thousands and thousands of data were gathered to create population statistics and statistics on agriculture, industry and trade. But [...] not much was done with it. Often, there was a real desire to present geographically relevant issues with the help of a map, but the authors lacked the knowledge to do so. In the end, the result of all these efforts was not more than a book full of tables, sometimes supplemented with a few ordinary geographical maps" (Koeman 1983, 244). So, there was no synergy; Somerhausen, Kuyper, and Mees remained lonely pioneers in the field of thematic cartography in the Netherlands. Their efforts remained largely unknown; their work was never copied (Koeman 1983, 247).

The negligence or inability of Dutch statisticians and Dutch cartographers to find and assist one another in the production of thematic maps is not a problem that only existed in the Netherlands. Elsewhere, in Britain and Germany, it took quite some time before the first examples of choropleth mapping were replicated. And even in France, where Dupin's map had been reproduced by Guerry in 1833, the real breakthrough of choropleth mapping came after 1860, probably because Dupin was no geographer by origin and therefore an outsider to the geographical scientific community (Palsky 2004, 593).

From these examples, a conclusion can be drawn that the first published choropleth maps had been actually created by outsiders, by people who were neither cartographers nor statisticians. They remained solitary scientists whose work was neither copied nor reproduced. Only much later, when a new generation of statisticians and cartographers started to work within the departments of geography, it became possible to exchange knowledge about statistics, data, and technology and

[7]Molhuysen and Blok (1914, 723). His 1857 maps were the first he ever made; later, he started to propagate the use of maps in geography and became the founding father of the prestigious *Koninklijk Nederlands Aardrijkskundig Genootschap*.

to produce thematic maps. Only when outsiders were admitted and statistics and cartography were included into the domain of scientific geography did the golden age truly emerge. Scientific inclusion proved to be the fourth necessary condition to inaugurate the golden age of thematic cartography.

From this, we may conclude that it is not only statistics, data, and technology that are needed to create a perfect storm necessary to produce the golden age of thematic cartography. There is needed one more ingredient: scientific inclusion.

3.5 Prerequisites for a Golden Age of Historical GIS

The history of choropleth mapping may be an exemplary predecessor to the history of historical GIS or, anyway, lead to a number of questions related to historical GIS. For instance, for thematic cartography to emerge, four conditions: statistics, data, technology, and scientific inclusion, were needed. Does this mean that something similar, for example, data, GIS technology, and scientific inclusion are needed in order to instigate a golden age of historical GIS? If so, is a perfect storm to enter into a historical GIS golden age coalescing at this very moment? Or are we still in a preliminary stage in which mere outsiders are working with historical GIS, with one or more of the prerequisites, be it data, technology, or scientific inclusion, lacking?

Let us answer the last question first: It is obvious that historical GIS exists and that it has existed for a considerable amount of time. The first examples of historical GIS date from the work of Jan Hartmann in the Netherlands. This medievalist devised a GIS of its own between 1980 and 1985. All programming (in FORTRAN) was accomplished by himself. Hartmann created his GIS in order to analyze medieval land reclamations in the southern part of the province of Limburg (Hartmann 1986; see Fig. 3.2). Just like Dupin had been, in combining the disciplines of history, geography, and information science, he was able to conduct pioneering work with his own devised historical GIS. And, as in the case of Dupin, nobody followed his footsteps.

Since the 1980s, technical progress has led to GIS software that is very user friendly; quantitative, digitized data have become abundantly available; historical GIS is used at many universities around the world. So what is stopping us from claiming this era as a golden age of historical GIS? I think that if we look more precisely into the necessary prerequisites for such a golden age, a number of problems emerge that prevent the necessary conditions to be met.

1. *Data*: Large digitized data sets are available, many of which are easily within reach through the Internet. Only a small amount of the data is organized in such a way that it could be geo-referenced synthesizing location to time more or less automatically with location-related meta-information, thus simplifying the plotting of the data onto maps. Digitized maps are much less available. There are thousands of historical maps in the archives, but only a small number have been digitized; a smaller number has been geo-referenced and an even

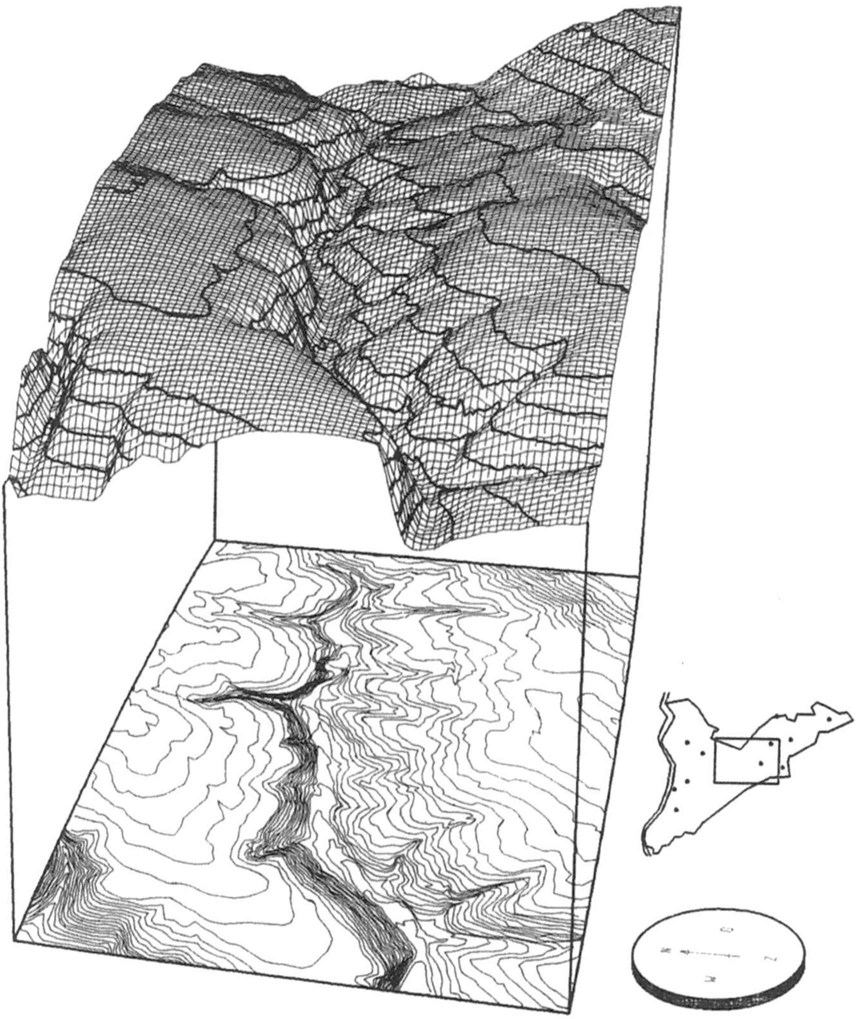

Fig. 3.2 Among the first historical GIS examples in the Netherlands, Jan Hartman's FORTRAN programmed analysis of medieval land reclamations in the province of Limburg (With permission from Jan Hartmann)

smaller number have been vectorized. Without geo-referenced historical maps and without location-stamped data, research based on historical GIS will take too much time and too much effort to compete with "traditional" historical research. Only when a relevant number of maps have been digitized, vectorized, and geo-referenced, and only when location-stamped data have become abundantly available, has historical GIS a chance of taking off.

2. *GIS technology*: From Jan Hartmann's FORTRAN programming, via software packages like AutoCAD to standard software like ArcGIS, GIS technology has

created not only very user-friendly software but also a wide variety of GIS tools, both of which have made the application of historical GIS a lot easier and well within reach of nonspecialists, such as historians. However, standard GIS software still is not able to tackle some of the problems that are so typical for historical GIS: temporal changes, for example, or the handling of missing values, ambiguities, errors, and inconsistencies in historical data. There are GIS tools to solve such problems, but they are not included in standard GIS software and therefore not widely available. As a result, historians may be disappointed by the possibilities historical GIS offers.

3. *Scientific inclusion*: Hardly any of the research done within the domain of historical GIS has been carried out by historians. Most contributions are from geographers, economists, archivists, city planners, earth scientists, environmentalists, sociologists, and so on. GIS-related historical research seems to be restricted to a scientific domain outside traditional historical departments. Historical GIS has not been accepted within traditional historical departments. Therefore, scientific inclusion, at least within the departments of history, is lacking.

One can conclude that there are lacunae in all three domains. All three lack the proper conditions for the perfect storm to produce a golden age of historical GIS. But, as has been argued previously, it is academic culture within historical science that poses the biggest barrier for historical GIS in the years to come. All over the world universities are very reluctant to reserve room within their curricula for teaching research methods and techniques that go beyond the methods of traditional historical science. As a consequence, hardly any history student will have the chance to receive a proper training in the use of historical GIS (Boonstra 2009a,b).

3.6 Conclusion: No Storm Will Rise, No Golden Age Will Dawn

The first choropleth map in the world, the *Carte figurative de l'instruction populaire de la France*, was made in 1826 by Pierre Charles Dupin. His map was copied by Hartog Somerhausen one year later, but it took decades before choropleth mapping became an accepted tool for visualizing regional variations. The reason was not a lack of statistical skills, data, or technological achievement; it was because Dupin and Somerhausen were outsiders to the discipline of geography, where for a long time thematic mapping was not acknowledged for what it was worth: an innovative way of visualizing quantitative data. As a result, between 1826 and 1860, choropleth mapping was hardly conducted, and if so, it was conducted by outsiders. Only when geographers, skilled both in statistics and cartographical methods, recognized the value of thematic mapping and then when these skills were introduced and taught within the departments of geography did thematic cartography become a standard tool.

This chapter argues that something similar is happening at the moment with historical GIS. Although the first uses of GIS within historical science and undertaken by historians date back to the 1980s, historical GIS still is not part of mainstream historical research. And as long as history departments remain reluctant to add the teaching of historical GIS to their curriculum, a perfect storm for historical GIS will not rise, let alone inaugurate a golden age of historical GIS. Frankly speaking, the mere fact that we talk about "historical GIS" instead of "GIS-based history" is a clear indication that no golden age of historical GIS has dawned—yet.

References

Anonymous (1827) Rapport sur les écoles du Royaume adressé aux Etats-Généraux per le Ministère de l'Instruction. Weissenbruch, Bruxelles

Anonymous (1842) Verslagen nopens de Staat der Hooge, Middelbare en Lagere Scholen in het Koninkrijk der Nederlanden, gedaan aan de Staten-Generaal 1816–1840s. Algemeene Lands Drukkerij, Gravenhage

Beniger J, Robyn D (1978) Quantitative graphics in statistics: a brief history. Am Stat 32:1–11

Boonstra O (2009a) De daling van het analfabetisme. Een regionale analyse over de periode 1775–1900. In: Boonstra O, Schuurman A (eds) Tijd en ruimte. Nieuwe toepassingen van GIS in de alfawetenschappen, Matrijs, Utrecht

Boonstra O (2009b) No place in history – Geo-ICT and historical science. In: Scholten H, van de Velde R, van Manen N (eds) Geospatial technology and the role of location in science. Springer, Berlin/Heidelberg

Dupin C (1826) Carte figurative de l'instruction populaire de la France. Bachelier, Paris

Frederiks J, van den Branden FJ (1888–1891) Biographisch woordenboek der Noord- en Zuidnederlandsche letterkunde. L.J. Veen, Amsterdam

Friendly M (2007) A.-M. Guerry's moral statistics of France: challenges for multivariable spatial analysis. Stat Sci 22(3):368–399

Friendly M (2008) The golden age of statistical graphics. Stat Sci 23(4):502–535

Friendly M (2011) Milestones in the history of thematic cartography, statistical graphics, and data visualization. http://www.math.yorku.ca/SCS/Gallery/milestone/milestone.pdf. Accessed 23 Sept 2011

Hartmann J (1986) De reconstructie van een middeleeuws landschap: nederzettingsgeschiedenis en instellingen van de heerlijkheden Eijsden en Breust bij Maastricht (10e–19e eeuw). Van Gorcum, Assen

Kasper-Holtkotte C (2003) Im Westen Neues: Migration und ihre Folgen: deutsche Juden als Pioniere jüdischen Lebens in Belgien, 18./19. Jahrhundert. Brill, Leiden

Klep P, Stamhuis I (eds) (2002) The statistical mind in a pre-statistical era: The Netherlands 1750–1850. Aksant, Amsterdam

Koeman C (1983) Geschiedenis van de kartografie van Nederland: zes eeuwen land – en zeekaarten en stadsplattegronden. Canaletto, Alphen aan den Rijn

Kuyper J (1857) Geographische und statistische Darstellungen der Niederlande. Petermann's Mitteilungen p Tafel 82

Mees G (1840) Redevoering over het belang om der studerende Jongelingschap lust in te boezemen tot eene grondige beoefening der bronnen van onze Geschiedenis, opdat zij zich bekwamen moge om de geschiedkundige wetenschap in ons Vaderland te helpen bevorderen. Ter aanvaarding van het Professoraat in de Nederlandsche Taal en Letteren en Vaderlandsche Geschiedenis aan de Doorluchtige School der Stad Deventer gehouden 9 April 1840. De Lange, Deventer

Mees G (1865) Historische Atlas van Noord-Nederland van XVI eeuw tot op heden. Verbruggen & Van Duijm, Rotterdam
Molhuysen C, Blok P (eds) (1914) Nieuw Nederlandsch biografisch woordenboek, vol 3. A.W. Sijthoff, Leiden
Palsky G (1996) Des chiffres et des cartes. Naissance et développement de la cartographie quantitative au XIXe siècle. CTHS, Paris
Palsky G (2004) Le calcul par l'oueil. In: Bord JP, Baduel R (eds) Les cartes de la connaissance. Karthala, Paris
Potter T (1986) The rise of statistical thinking, 1820–1900. Princeton University Press, Princeton
Quételet A (1827) Recherches sur la population, les naissances, les déces, les prisons, les d'pôts de mendicité, etc. dans le Royaume des Pays-Bas. Tarlier, Bruxelles
Randeraad N (2002) The Dutch path to statistics, 1815–1830. In: Klep P, Stamhuis I (eds) (2002) The statistical mind in a pre-statistical era: The Netherlands 1750–1850. Aksant, Amsterdam, pp 99–123
Schwebber L (2006) Disciplining statistics. Demography and vital statistics in France and England 1830–1885. Duke University Press, Durham
Somerhausen H (1827) Carte figurative de l'instruction populaire des Pays-Bas [map ca. 1:1.600.000]. [S.I.] : [s.n.], [ca. 1828]; University Library Utrecht, Ackersdijck 164
Wallis H (1989) 19-eeuwse thematische kartografie. Caert-Thresoor 8(2):46–52
Weijermans J (2010) Twee prijsvragen, twee strategieën. Jan Frans Willems, Prudens Van Duyse en het Brusselse genootschap Concordia (1818–1830). In: Bemong N (ed) Naties in een spanningsveld : tegenstrijdige bewegingen in de identiteitsvorming in negentiende-eeuws Vlaanderen en Nederland, Verloren, Hilversum, pp 79–96
Woodward D (1975) Five centuries of map printing. University of Chicago Press, Chicago
Woolf S (1989) Statistics and the modern state. Comp Stud Soc Hist 31(3):588–604

Chapter 4
(In)Digitizing Cáuigú Historical Geographies: Technoscience as a Postcolonial Discourse

Mark H. Palmer

4.1 Introduction

In this chapter, I will consider how combining postcolonial discourse and actor network theory can inform emerging research on historical GIS (HGIS) and more specifically, the processes of co-constructing historical geographies of North American Indians. First, I will frame my consideration within the body of literature on critical cartography and GIS, especially as it pertains to GIS and indigenous people. Second, I will introduce a new term called indigital as a form of postcolonial discourse. Indigital HGIS is a form of postcolonial hybridity and the process of bringing two things together to create a third. Next, I will introduce the three primary theoretical positions within science and technology studies (STS) including the technological-shaping thesis, the social construction of technology, and the co-construction of technology and society thesis (Brey 2003) as an amalgamation of ideas. I will primarily focus upon concepts found in actor network theory: intermediaries, actants, translations, obligatory points of passage, and the center of calculation. Fourth, I present an emerging indigital Kiowa HGIS including the translation of problems, interests, enrollment, and mobilization of the project. Fifth, an ANT concept that can help describe historical, inscriptive processes is considered as the center of calculation. And finally, I will provide some concluding remarks.

4.2 Critical Cartography and GIS

Critical cartography and GIS is a social theoretical critique of maps and GIS, questioning how cartographic/GIS representations come to be, who creates such

M.H. Palmer (✉)
Department of Geography, University of Missouri, Columbia, MO, USA
e-mail: palmermh@missouri.edu

A. von Lünen and C. Travis (eds.), *History and GIS: Epistemologies, Considerations and Reflections*, DOI 10.1007/978-94-007-5009-8_4,

representations, what purposes are they made for, and who do maps and GIS empower and marginalize (Crampton 2010, 39). One of the early scholars influencing the intellectual tenets of critical cartography and GIS research was Brian Harley. Harley's work drew upon insights from Foucault and Derrida to produce manuscripts on maps, knowledge/power, and deconstruction of the maps as texts (Harley 1989). Harley's intellectual contributions within the history of cartography acted as a precursor to the GIS and society debates that occurred during the 1990s. Much of the early debate "pitted a social theoretic perspective of GIS" as technology that would bring scientific credibility to the discipline of geography (Sheppard et al. 1999, 801). A second wave of critiques involved a greater level of collaboration between human geographers and advocates of GIS through publications like *Ground Truth* edited by Pickles (1995) and a special issue of *Cartography and Geographic Information Systems* on the social implication of GIS (Schuurman 2000). Issues pertaining to representations, research, ethics, geodemographics of marketing, indigenous epistemologies, and research agendas entered the discourse on geographic information technologies in general.[1] Most recently, a third wave of research diverged into multiple areas of inquiry including the social construction of GIS,[2] public participation research agendas and case studies,[3] and emerging research on volunteered geographic information (Goodchild 2007). A more detailed analysis of the GIS and society debate can be found in Schuurman's "Trouble in the Heartland" (Schuurman 2000). The remainder of this chapter is dedicated to opening up new map and GIS territories for discussing American Indian historical geographies and the creation of alternative representations in GIS and maps.

4.2.1 GIS and Indigenous People

The critique of cartography and GIS within indigenous communities is both spatial and historical. Melinda Laituri recently argued that indigenous mapping and GIS must be situated within the contexts of the Columbian exchange, industrial revolution, and globalization, which shaped and shapes the politics of representation, access, public participation, indigenous cyberspace, and prospects and new directions for research (Laituri 2011). American Indian geographies and cartographies have been appropriated and influenced by cartographic encounters with Europeans over the past 400 years (Lewis 1998). Several scholars have documented some of the encounters found in archives (Lewis 1998; Warhus 1997). More recently, mapping and GIS practices and processes have revealed the centralized, top-down GIS networks emanating from the United States Bureau of Indian Affairs (Palmer 2009, 2012b). Some American Indian tribal governments and groups have developed

[1]Pickles (1995), Curry (1998), Rundstrom (1995) and Sheppard (1995).

[2]Harvey and Chrisman (1998) and Harvey (2001).

[3]Harris and Weiner (1998), Craig et al. (2002) and Sieber (2006).

historical GIS for land tenure, trails, estuaries, tidal channels, and landscape change (Sappington 2008).

Key issues shaping knowledge of GIS and indigenous people focused on the epistemological compatibility of GIS technologies with indigenous geographic knowledge. Robert Rundstrom argued that the geographic relationships constructed by GIS were not compatible with indigenous geographic knowledge (Rundstrom 1995). Indigenous cultures "emphasize oral communications and other performance-based modes (e.g., dance, painting) in transmitting all sorts of meaningful information" (Rundstrom 1995, 51). Some scholars argued that different social groups (indigenous people) employ different ways of knowing (different ways of reasoning about and making sense of the world) which need not be consistent with those underlying conventional databases and GIS software (Sheppard et al. 1999, 810). Increasingly, scholars and indigenous informants have experimented with indigenous knowledge and GIS to gain a better understanding of how both systems are similar as well as different. For example, Robins (2003) work in the GIS of local knowledge concluded that for both local indigenous people and expert managers, the cultural meaning of landscapes was dependent on their roles in regional resource production and resource politics. There have been recent efforts to formalize indigenous knowledge through conceptual and logical ontologies for the Cree Indians in Canada (Sieber and Wellen 2008). Other studies investigated the similarities and differences in conceptualizing landscape features among the Yindjibarndi in Australia and the Navajo in the southwestern United States. Mark and Turk (2003, 16) found "several important differences between the Yindjibarndi and English languages regarding conceptualizations, the ontology, of water in the landscape." Furthermore, Yindjibarndi speakers used different words for the same physical feature depending on the informant's point of view (Mark and Turk 2003). Sieber argued that "[t]he challenge in preserving knowledge may be to fit the technology to the knowledge instead of the converse" (Sieber and Wellen 2008, 2). However, the separation between Western and indigenous is false and more uncertain than this literature presents.

Efforts to break down the dichotomy between indigenous and scientific mapping and GIS have emerged through map and GIS design. During the twentieth century, there has been multiple "kinds of fusion, interbreeding, and boundary crossing in mapping and GIS among indigenous peoples of the world [...] it's the new attitudes, visions, and radical philosophies of the counter-mappers that are really taking maps and mapmaking [GIS] in a whole new direction" (Wood 2010, 111). Margaret Pearce focused her research on design and innovative mapping language to represent historical journeys, landscapes, and place.[4] It has been observed that cartography as a process "connects oral, written, performative, and experimental modes of mapping as a means to transmit situated indigenous cultural knowledge from one generation to the next" (Pearce and Louis 2008, 110). There is also a need to include marginalized voices in GIS, to represent their own histories

---

[4]Pearce (2008), Pearce and Louis (2008) and Pearce and Hermann (2010).

and ideas, and put native languages into practice. Kiowa stories and language have successfully converged with a transportation network GIS as a model and alternative to English-only systems existing in American Indian communities (Palmer 2012a).

Alternative representations and the use of maps and GIS by indigenous communities and scholars are an established discourse in research on critical cartography and GIS. The histories of indigenous peoples, in the Americas, are situated within the historical contexts of the Columbian exchange, industrial revolution, and globalization. One of the tenets of globalization is the exchange of ideas, materials, and knowledge between the peoples of the world. As a result of this new kind of dialogue, histories and geographies shift and vary from region to region, community to community, and from person to person. The amalgamation of peoples, their knowledge systems, and modern technologies is already underway. HGIS must be pliable enough to incorporate what has been described above as alternative representations that are simultaneously spatial and temporal.

4.2.2 Indigital(ization) as a Postcolonial Discourse

Postcolonialism is a critical approach to analyzing the impacts of colonialism, but also seeks to present alternative views of the world (Sharp 2009, 4). Navigating through postcolonial territories of the mind is not an easy task as the roads are littered with linguistic casualties, broken families, distrust, anger, and regrets. Indigenous histories and geographies are potential mine fields strewn with contention and alternative narratives. There is an art to tiptoeing through the mine fields laid by colonialism. Shifting and maneuvering through the field often leads to the creation of something new, acknowledging the benefits of past and present encounters and exchanges, and openly discussing the failures of colonial projects. I take the position that "postcolonialism is also a more positive project which seeks to recover alternative ways of knowing and understanding" the world, geographies, maps, and GIS (Sharp 2009, 5).

More voices speaking "pushes the discourse on to a different plane and into a recreative space of radical openness where both development and social justice can be revisioned together, along with their histories and geographies, not as an either/or choice but in the limitless expansion of the both/also [...] [a]s Thirdspace, the critique does not come to a full stop at this third way; nor is it a simple additive or 'in-between' positioning that can be marked with dogmatic assurity. It is instead an invitation to continuous deconstruction and reconstruction, to the constant effort to move beyond the established limits of our understanding of the world" (Soja 1996, 126). Soja's vision of a "Thirdspace" is the philosophical basis for what I refer to as the "indigitalization of HGIS".

The term indigital is a "neologism that describes an amalgamation of indigenous, scientific, and digital technological knowledge systems; characterized as fragmentary, contradictory and full of uncertainties" (Palmer 2012a). This term is an amalgamation of postcolonial, poststructuralists, and humanist ideas as they relate to information technologies, science, and more specifically GIS. "Indigitalization" is the process of connecting indigenous and technologies through digital convergence. Indigitalization enhances native science ideas (Cajete 2000). Indigital thought does not subscribe to the idea of the all-knowing single author or the all-knowing indigenous shaman. Rather, an indigital process is analogous to an intertextual canvass setup for the interplay of many kinds of knowledge and emerging concepts in an age of increasing connectedness and interactions.

If indigital is to emerge as a concept, it needs to remain open, allowing for creative input. However, there are a few guiding tenets that are needed to maintain openness. Indigital processes are the following:

1. The mixing of indigenous, scientific, and technological knowledge systems;
2. fragmented, contradictory, and uncertain;
3. spaces for combining indigenous, technical, and scientific systems because they are open, dynamic, and ever-changing;
4. the mixing of two things to create a new third;
5. reciprocal;
6. everywhere in the world and the same time conceptually distant from the reality of many indigenous groups on a global scale.[5]

Indigitalization encourages the breakdown of dichotomies between Western and indigenous knowledge systems like language, science, local knowledge, histories, and everyday lived experiences. Perhaps indigenous and scientific knowledge systems, like an HGIS, should not be seen as competing paradigms but rather viewed as complementary systems having strengths and weaknesses. Indigitalization of histories includes the development of "digital songlines" in Australia, combining multimedia curriculum with oral traditions, creating a Dena'ina language archive, development of kinship and genealogical databases, and incorporating Kiowa stories into a network GIS.[6] "The multimedia capabilities, storage capacity and communication tools offered by information technology provide new opportunities to revitalize indigenous cultures and languages, and to repatriate material back to communities from national cultural institutions. In particular the graphical, video and audio facilities of multimedia speak directly to cultures which are principally rooted in spoken language, music, dance, ceremony and visual forms of artistic expression" (Dyson et al. 2007). With this said, another dichotomy is encountered: what drives indigitalization of HGIS, technology, or society?

---

[5]Palmer (2012a).

[6]Leavy (2007), Inglebert et al. (2007), Holton et al. (2007), Sugito and Kubota (2007) and Palmer (2012a).

4.3 The Science, Technology, and Society Amalgamation

The indigitalization of historical geographic information systems (HGIS) is a technical process that can handle, manipulate, and represent historical information. This approach is dependent upon the three components of GIS including the presence of attributes, the location of things, and time. Attributes are characteristics of data being represented in the database that can be mined or queried to find useful spatial patterns. HGIS are also able to incorporate digitized oral narratives and histories as attributes tied to some geographic location. Locations are places on the Earth's surface. The spatial resolution of places can range from precise latitude and longitude coordinates to street intersections to addresses to place-names. In addition to text, digitized symbolic images like calendar pictorials, nameglyphs, pictographs, or petroglyphs could be digitized and geospatial information integrated into the image itself. In other words, interested persons would have to go through indigenous symbols to access geospatial information about the image or narratives, putting indigenous knowledge into practice with digital environments. Time is not well represented in GIS, although various visualization techniques (movies, time-series maps) are addressing the problem of representing time. One could argue that the indigization of HGIS is dependent upon the capabilities and convergent characteristics of computerized digital technologies.

Yet, viewing HGIS as a purely technical process is a form of technological determinism. Although the strong position on technological determinism is generally dismissed by scholars, weaker forms of the argument are useful. Technology may not determine the direction of society, but technology does transform society. The technological-shaping thesis claims that society is strongly shaped by technology. This means that there are multiple "functions, meanings, and effects" that always accompany the development and implementation of technology (Brey 2003, 52). Technology has political consequences (Winner 1980), can contain gender biases (Bray 1997), and shapes the behavior of users (Latour 1991; Law 1991). The history of cartography documents the ways in which maps contributed to the colonization indigenous peoples in North America (Lewis 1998; Warhus 1997). The examples above represent weak forms of technological determinism. As information technologies play important roles in our work as professional geographers, historians, or members of indigenous communities, our questions run the risk of being guided by what HGIS can or cannot do.

An opposing argument positions HGIS as purely social processes. The social construction of technology thesis claims that technological development can be explained solely through an analysis of social processes that include identifying social groups, interpretive flexibility regarding design, and technological stability (Bijker et al. 1987). Furthermore, the "claim that technological change can be entirely analyzed as a result of processes of social negotiation and interpretation, and that the properties of technologies are not objective, but are effectively read into the technologies by social groups" (Brey 2003, 51). Social groups interpret technologies in different ways, and ultimately shape an artifact's development

(Bijker et al. 1987). However, social construction of technology theorists are criticized for giving too much explanatory power to social groups and social processes in the development of technology while neglecting the influence of nature or other forms of technology on development. The social construction of technology has contributed to the literature on GIS and society.[7] These studies include the social construction of boundary objects and GIS technologies (Harvey and Chrisman 1998; Harvey 2000). Both the social and technological determinist models are partially true. Yet, digital geospatial technologies do not emanate from all societies, and not all social groups have a hand in shaping technologies. How might we deconstruct and address the have/have not dichotomy? A third STS theoretical position called actor network theory challenges these ideas by arguing that society and technology co-construct one another (Brey 2003).

The STS approach to the co-construction of society and technology is actor network theory (ANT). Actor network theory emerged from the research of Latour (1987), Callon (1991), and Law (1999), among others. Actor network theory is a conceptual framework for "exploring collective sociotechnical processes," describing and explaining the "relationships between people, institutions, and artifacts connected by agreements and exchanges" (Harvey 2001, 30). ANT is also a method for tracing relationships, agreements, and exchanges. General symmetry between humans and nonhuman means that no separation or dichotomy exists between society and materials. ANT sees relationships as a coherent whole (Latour 2005, 76). Dichotomies like technology/society or GIS/society are false. They are one and the same. Through relationships and agreements, action is achieved at a distance. Geographers use ANT as a framework to understand the interactions between GIS and society, including the interactions between texts, people, money, technology, and control in the implementation of GIS.[8] Furthermore, perhaps "the best use of ANT for investigating GIS may be to continue exposing the social interactions behind GIS operations so practitioners, managers, theorists and researchers will be more sensitive to building stable GIS actor networks" (Martin 2000, 735). Actor networks are built of materials and ideas. Indigital HGIS actor networks are evolving within a playful interaction between many actants and their materials; indigital HGIS is in the making. In the sections to follow, I will lay out some ANT concepts and demonstrate how they inform the historical GIS debate (HGIS approach).

4.3.1 Intermediaries

Actor networks consist of connections between materials called intermediaries and actors. An intermediary "is anything passing between actors which define the relationship between them [...] examples of intermediaries includes scientific arti-

[7]Harvey and Chrisman (1998), Harvey (2000) and Schuurman (2000).

[8]Palmer and Hanney (2010), Palmer (2009) and Martin (2000).

cles, computer software, disciplined human bodies, technical artifacts, instruments, contracts and money" (Callon 1991, 134). The sheer number of intermediaries is overwhelming.

Intermediaries can be limited to text inscriptions, technical artifacts, human beings, and all forms of money—funding (Callon 1991). Texts include influential publications in peer-reviewed journals or books like the special editions in *Social Science History* (vol. 24: 3), *Historical Geography* (vol. 33), and review essays in *Progress in Human Geography*. Research papers and books as intermediaries are the work of scholars arguing for increased use of GIS in history and the humanities. Cutting-edge work becomes the domain of "young scholars whose dissertation research was grounded in the use of GIS as a core method of analysis. They represent the leading edge of what seems certain to become a new generation of historical scholarship built with GIS" (Knowles 2005, 7). The flow of money is important for the propagation of GIS. For instance "in the UK, North America, China, and Western Europe there have been very significant sums of money invested by research funders in the development of national historical GIS. These projects, by their very nature, are large and consequently costly, with budgets that may run into the millions" (Gregory and Ell 2007, 199). New geospatial technologies include increasingly user friendly interfaces, standardized data, and the digitization of archival documents. The justification here is that "GIS software is now relatively cheap, reasonably easy to use, and runs on standard PCs. Desktop GIS is thus well established and mature, and is likely to change little in the near future. Scholars can therefore, proceed to use GIS tools in the confidence that the software is robust, well supported, and will not be transitory" (Gregory and Ell 2007, 200).

Intermediaries help define the roles of human and non-human actors (actants) in the network (Murdoch 1995). Depending on what your intermediary data source might be, historians rely upon quantitative, qualitative, or mixed methods. GIS is well equipped to receive, manipulate, and represent quantitative data sources like census, crime reports, or disease surveillance. It is recognized that the "spatial component of data in a GIS [software] are fundamentally quantitative, as they consist of co-ordinate pairs-numbers that can be manipulated mathematically" (Gregory and Ell 2007, 161). Georeferenced qualitative data sources like oral histories, photographs, video images, and audio files can also be applied within a GIS framework. Meaning is inscribed within intermediaries such as maps, database attribute tables, census data, and natural resource assessments. Words within documents can be traced to other documents.

Historical geographies of indigenous people engage with gridded and non-gridded geographic data sources or intermediaries. One of the first things that researchers of indigenous people must consider is whether or not we (they) are going to continue to write and create state geography as related to the experience of North American Indians and other indigenous peoples. State geographies are based on a grid system like administrative units, cadastral maps, township and range system, national maps GIS data, census data, and public land office maps and surveys. Each of these intermediaries are very valuable and can provide much information about changing populations, voting patterns, the evolution of economies, and historical cartography of scientific mapping in the United States. Furthermore grid data sources seem to be the most compatible with computer-based geographic information systems as a software package. Perhaps the most important factor in this compatibility is the all-important geo-reference information that can be found on historical paper maps and easily integrated into GIS.

State geographic intermediaries like reservation maps, allotment maps, and state sponsored ethnographies are highly useful for researching North American Indian historical geographies. Reservation policies, surveys, and maps are the cornerstone of America's internal colonial policies toward American Indians (Palmer and Rundstrom 2013). Reservation surveyors calculated and measured the locations of boundaries precisely. In fact, there are a multitude of archival sources that describe, spatially, nineteenth-century reservations. Much more elusive are allotment maps created primarily during the twentieth century or the post-Dawes Act of 1892 leading to the allotment of reservations. Unlike the reservation maps, one cannot simply search for and find allotment maps on the Internet. Researchers must physically visit the Bureau of Indian Affairs agency and area offices or the national archives or state archives to find original allotment maps. Allotment geographies are checkerboard patterns of what might be described as islands of federal trust properties. Among the various reservation allotment maps are found archival documents pertaining to the old Indian agency or as mentioned before the Bureau of Indian Affairs. National Archives and the Smithsonian Institution in Washington DC hold documents directly connected to John Wesley Powell's bureau of ethnography field research projects that hold much information on the classification of American Indian tribes and their cultures. Reservation maps, allotment maps, archival documents, and US Bureau of Ethnography data are all examples of state or grid geographies of American Indian territories. But there are other lesser known sources that emerge from American Indian communities that must be included in any GIS, if it is to be seen as legitimate within American Indian communities.

Some intermediaries associated with the indigenous historical geographies like oral histories, stories, pictorial calendars, and rock art are not completely compatible with grid space. Oral histories are held and shared by knowledgeable people and elders from indigenous communities. Some stories are mythical, others are historical, and the most personal histories are family stories that are rarely shared. Surprisingly, many oral histories are found on audiotapes located in many libraries and archives. The Doris Duke Oral History Collections are found at the University of Oklahoma, for example. Within oral histories are stories about people and places. Places are significant to our understanding of indigenous historical geographies. Mythical, historical, or family stories are equally important for understanding the historical now. Some American Indian tribes kept cyclical, seasonal calendars. Many of the calendars contain pictorial images that in themselves hold much information and meaning about places. Related to the pictorials are many places that feature rock art. Important questions associated with these intermediaries include understanding how to interface GIS and indigenous knowledge systems. Oftentimes, this requires interacting directly with the keepers of the knowledge.

4.3.2 Actants

Actants are human and nonhuman actors that author intermediaries. Actants combine, degrade, and compute intermediaries to create new intermediaries (Callon 1991, 141). Inscriptions are the way actants write (Curry 1998, 40). Actants author books and peer-reviewed journal articles that focus on all kinds of content and ways in which the content is presented. Historiography is a form of inscription or a way of writing or arguing. Actants are not preordained with characteristics. One could trace and read the intermediaries and associated networks to gain more knowledge about actants. A content analysis of recent work on history and GIS could provide us with an idea of how strong GIS actor networks are within history, who the actants are, and what intermediaries they prefer. Actants, as mediators, transform and put intermediaries into motion.

Networks are groups of interconnected actants and intermediaries. Networks can be strong and durable or initially, weak and evolving/devolving over long distances (Latour 1987, 127). The success of GIS networks within the discipline of history or the humanities will depend upon the length, durability, and strength of networks. From the geography side, historical GIS scholars have gained footing in the discipline through publications like *Progress in Human Geography*, several special issues within journals, and a handful of books. By following the history of GIS networks, it is possible to locate and describe additional actants and trace interconnected intermediaries (Callon 1991, 142). A network works as "a coordinated set of heterogeneous [actants] which interact more or less successfully to develop, produce, distribute and diffuse methods for generating goods and services" (Callon 1991, 133). The amount of heterogeneity varies from network to network. Actants can actually define one another through their interaction, including the authoring and circulation of intermediaries from one institution to another (Callon 1991, 135). A co-construction of humans and nonhumans takes place, and negotiations and agreements must be presented and accepted. Actants need to be enrolled or allied closely with other actants and intermediaries to maintain stability. Throughout the process, scientists become powerful by enrolling participants to help them build durable networks (Murdoch 2006, 74–75). The process of translations aligns actants, stabilizing or destabilizing networks. Indigenous communities know their histories, have knowledge of materials but are not remotely interested in advancing the field of HGIS. Geographers and linguist must present communities with problems that may interest people in enrolling and mobilizing bottom-up HGIS actor networks. An indigital HGIS is in the making. However, the first step in exploring such an idea is negotiating with communities of actants.

4.3.3 Translation

Translations are the goals, objectives, interests, projects, and desires negotiated between actants (Callon 1991, 143). As a network builder, one actant must convince other actants that their goals and interests should be aligned and converge (Callon 1991, 143–145). Some historical geographers select geospatial technologies, funding agencies, funding opportunities, management plans, or materials to help them achieve their goals of creating historical geographic information systems. Actants negotiating a translation argue something like, "You have a problem that I can solve, but you have to follow my instructions and guidelines precisely." Historical geographers, technical specialists, scientists, engineers, and others "speak in the name of new allies that they have shaped and enrolled; representatives among other representatives, they add these unexpected resources to tip the balance of the force in their favor" (Latour 1987, 259). Translations are embedded within intermediaries like texts, peer-reviewed papers on research methodologies, research agreements with informants, classroom curriculum, GIS software manuals, project guidelines, roundtable discussions, conference proceedings, or various newsletters (Callon 1991, 135–136).

Translation makes two different actants equivalent. This requires an alignment of intermediaries like statistical methods, GIS data standards, or funding. Translation is "an idea that suggests that if scientific networks are to be extended through space and time, then actors of differing (natural and social) types must be 'interested' into the network—that is, their goals must somehow be aligned with those of the scientists" (Murdoch 2006, 62). As mentioned above, actants will accept those translations that will help them maintain their networks and achieve their goals and objectives. Actants define problems, gateways, interests, enlist other actors, and set their plans into motion.

The first step in the process of translation is "problemization." Problemization defines a problem or a controversy that is in need of a solution. In problemization, actants seek to present themselves as indispensable to the other actants (Callon 1986, 69). One way in which HGIS could solve a problem and build strong networks would be through state or federally funded national historical geospatial databases. Potential actants within the network have to pass through them to access resources including grant funding, technical assistance, data, education materials, or licensing agreements with contractual obligations and professional responsibilities. Indigenous geographers go through tribal councils, elders, or clan mothers to secure permission to conduct ethnographic research within American Indian communities. Tribal councils, elders, and clan mothers are obligatory points of passage or gatekeepers. In other words, obligatory points of passage control resources and are partially responsible for the success of the actor network (Callon 1986, 70–71). During problemization and identification of the obligatory point of passage, all the major actants are presented and described.

4.4 The Problemization of an Indigital HGIS

One problem that linguists and geographers are proposing is how indigenous languages can be put into practice through the use of historical maps and GIS. What roles can digital geospatial technologies play in the overall project? At the preliminary stages of investigation, we find that GIS is a good organizer of indigenous spatially referenced oral histories and pictorial images. Coordinating the efforts and roles of actants, however, is important for stabilizing indigital HGIS networks. In the paragraphs below, I will describe my personal experiences co-constructing an indigital Kiowa HGIS with academics and the Kiowa community. Kiowas are a North American Indian group who reside primarily in southwestern Oklahoma, USA. As (Greene 2009, 6) wrote,

> The Kiowa are a Southern Plains people, many of whom still live in southwestern Oklahoma on lands that were once part of the Kiowa reservation. The reservation, shared with the Comanche and Apache, was established in the mid-nineteenth century, its boundaries defined by the 1867 Medicine Lodge Treaty with the U.S. government. The reservation was part of a much wider territory that the Kiowa previously claimed as their own. Their earliest oral traditions record that they once lived near the headwaters of the Yellowstone River, in the foothills of the Rocky Mountains in what is now Montana. Sometime during the eighteenth century they moved eastward onto the open plains.

Several actants involved include academics at research institutions, grassroots linguists, community elders, and GIS. Academic actants include geographers, linguists, historians, and anthropologists who publish articles, books, and essays on indigenous people and their communities while at the same time, sharing knowledge and experiences with the communities they work with. There are several important sociotechnical issues to be addressed within indigenous communities including poverty, health, land rights, water rights, and endangered languages. Many within the academic community have formed relationships that support both their intellectual work as well as indigenous communities from which some of the academics actually come from. They work with public and private funding agencies that provide money but demand commitment to projects. Furthermore, there is a system of review in place. Academic intermediaries like peer-reviewed manuscripts, past work-related activities, facilities at universities, and the commitment of scholars to shared their knowledge all shape the academic's relationship with communities.

Mediators between academia and the Kiowa community are grassroots linguists. Kiowa grassroots linguists are groups of parents, grandparents, and their children who are putting their language into practice through bottom-up language immersion classes and curriculum. Members of this group tend to be urban or what might be termed "college town Indians." Some of these actants have received college degrees, graduate school education, others have taken some college courses, and generally, all have graduated from high school. They have access to scholars and educational resources and potentially funding those Kiowa community members who live in the allotment-reservation areas may not have access to. Grassroots linguists apply for public and private funding but often do not meet an agency's criteria for collaborating with academics and local communities and cannot show merit through peer-reviewed publications. Often, projects proposed by grassroots linguists are viewed by agencies as being too ambitious. On the flip side, grassroots linguist have a difficult time making material connections with Kiowas living in the allotment-reservation land area, because grassroots linguists are simultaneously insiders and outsiders. Issues of trust are vital among all the actants.

Important actants are Kiowa elders who hold the historical geographic information about their communities, migrations, and origins. They are self-taught, with no academic backgrounds or exposure to geographic theories. Oftentimes, the knowledge and information is held within the mind, within the imagination of the people. Community members are living repositories of this knowledge. Indigenous geographic knowledge and information is not held by one centralized individual. Rather, indigenous geographic knowledge and information is decentralized. Thus, understanding indigenous knowledge is essential to understanding the historical geography of many people and is often time-consuming to document. Elders within the Kiowa community are one example of obligatory points of passage or gatekeepers. There are important intermediaries that elders author like ceremonial symbols, prayers, origin stories, oral histories, paintings, or audio recordings. In 2012, a group of elders is working on a project to chronicle the history and the historical geography of the people using oral histories, stories, and digital maps as sources. The power or sphere of influence of elders is primarily local and based within Kiowa communities located in southwestern Oklahoma.

GIS are actants that generate digital representations like maps and author representations. Humans supply the data by entering information into attribute tables. But it is the inner workings of the computer program code and the software that produces a map image, computes proximity, density, slope, or quantifies continuous raster fields. These representations can contribute to shaping our understanding of historical geographies. GIS and maps are obligatory points of passage or gatekeepers, too. Commercially available GIS software has a wide breadth of functionality but is expensive to purchase and requires expertise to operate. HGIS, as an Internet-based system or mélange of digital technologies, is more accessible and its gates more permeable.

4.4.1 Interested Actants

When actants become interested in a problem, they connect with new actor networks at the expense of other competing actor networks. Interests are the actions "by which an entity attempts to impose or stabilize the identity of the other actors it defines through its problemization" (Callon 1986, 8). Interests are presented to potential actants who can help strengthen an evolving actor network.

Getting community members interested in an indigenous HGIS can prove to be difficult. For a strong indigital HGIS actor network to emerge within indigenous communities, one must consider the dynamics of the actants and intermediaries, the connections and disconnections. As a model, working with different actants is a bit like engaging with our solar system. For instance, the Sun has the greatest gravitational pull and aligns all the planets that also have a gravitational pull. An indigenous HGIS actor network is full of actants and intermediaries that do fluctuate a great deal and are quite unpredictable, there is much uncertainty associated with such a network.

An indigital HGIS actor network supports and promotes reciprocity over conquering and isolating. In other words, network builders should determine whether or not to compete with emerging community historical and linguistic networks. This is where the intellectual philosophies of the historical geographers, historians, GIScientists, and anthropologist can co-construct a network that builds up historical geographies as opposed to one that deconstructs and fragments. The intention of the Kiowa historical project is to be inclusive, decentralized, peer-to-peer, and open for continual revision over time. Presented on paper to the Kiowa communities, this is a very appealing proposal and initially creates some excitement among the people. However, there is a long history of distrust between the Kiowa's and historians, anthropologists, geographers, and linguists. Often, the Kiowa community will translate distrust as a dichotomy pitting "us" versus "them." For example, if you want to learn about the technical aspects of Kiowa history or language, go talk to academics. However, if you want to talk about the real history of the Kiowas, a history that is part of our oral traditions, then talk to us. Other community groups like the Kiowa Chronicles Inc. are working to develop methods for highlighting histories found in oral histories and stories so that they can represent a Kiowa historical geography. Problems get solved by working together, not against one another, and academic actants not only enroll other actants and intermediaries to support their publishing networks, but should also be willing to become assimilated into local, community process like elder councils, roundtable discussions, or ceremonies. This means giving up some power, a difficult thing to ask most actants to do.

4.4.2 Enrolling and Mobilizing Actants

When new allies are aligned, actor networks extend and may become stronger and act as one "unified whole" (Latour 1987, 131). As with the indigital HGIS, building strong alliances between GIS, historians, and scholars of humanities will have to involve trust and maintain strong network relations. Many things can strengthen or weaken the network. Enrollment into an actor network "does not imply, nor does it exclude pre-established roles. It designates the device by which a set of interrelated roles is defined and attributed to actors who accept them. [Interests] achieves [enrollment] if it is successful. To describe enrollment is thus to describe the groups of multilateral negotiations, trials of strength and tricks that accompany the [interests] and enable them to succeed" (Callon 1986, 10). People who are interested in an indigital Kiowa HGIS may not necessarily align with or enroll in the actor network. To date, academic linguists and geographers have been the primary actants interested in enrolling in the indigital Kiowa HGIS. Select Kiowa elders and their materials have also collaborated with Kiowa scholars on issues like storytelling (Palmer 2003) and emerging research on mapping storyscapes at Hueco Tanks, Texas. Grassroots linguists have not shown much interest in collaborating with scholars. In fact, the grassroots linguists resist academic assistance.

The final stage of the translation process is mobilization. Mobilization is "a set of methods used by the researchers to ensure that supposed spokesmen for various relevant collectivities were properly able to represent those collectivities and not betrayed by the later [...] translation is a process, never a completed accomplishment, and it may fail" (Callon 1986, 1). Mobilization of a single indigital Kiowa HGIS is problematic. Some issues that act against a prescribed or centralized HGIS are the decentralized nature of indigenous knowledge itself. Historical knowledge is held by multiple actants within the Kiowa community. Somehow, accessing and centralizing such a dynamic is highly unlikely and goes against what most Kiowas believe is their own, unique knowledge system. Somehow, academics and students with access to place-names, maps, and allotment areas must connect with community elders who maintain oral histories and stories locally, and Kiowa parents and grandparents who practice the Kiowa language with their children. This is no small task.

In summary, the strength of networks depends on how well actants translate the goals and objectives of their plans. Strong network alignment requires convergence of actants, intermediaries, and translations. Intermediaries are inscribed with rules, laws, standards, and meanings embedded by actants. Such entities are said to display strong co-ordination, which contributes to network stabilization. Successful translation processes lead to the stabilization of networks, and it is possible to trace these networks unproblematically. However, this alignment is not always the result of translation. Often there is controversy, conflict, or distrust resulting from disagreement among actors about the translation process. This in turn can lead to betrayal when actants refuse to accept the role assigned to them, and they align with another competing network. Historians, scholars of the humanities, and GIScientists must be able to talk with one another, and try not to revive old dichotomies that pit quantitative and qualitative research or the humanities versus science. In the same way, constructing an indigital Kiowa HGIS requires rethinking how geographic information is accumulated and how network building can involve mutual agreements among all actors. What do we do with the intermediaries collected in the field? How will intermediaries define all actants involved?

4.4.3 *Combining Materials and Ideas at the Center of Calculation*

Centers of calculation are the places where scientific maps, GIS, simulations, and models are constructed, refined, and used to act at a distance. Classic examples of centers of calculation are government bureaus (Palmer 2012b) or botanical gardens. Both places are interested in the world out there. However, scientists in the center are unsure of what lies on the periphery, on exotic islands, and the mysterious inhabitants of far-off places and their strange rituals. The centers send explorers, surveyors, anthropologists, geographers, and linguists out to the periphery to gather geographic information, and bring it back to laboratories.

Cycles of accumulation describe the reach of scientific institutions that go out to the periphery and collect geographic information. Information is brought back to the center and turned into scientific knowledge (Latour 1987, 120–121). Once back at the laboratory, the collected items can be studied, and more precise inscriptions are made in the form of maps, manuscripts, models, and simulations. Contemporary scientists construct facts within their laboratories, allowing officials and managers to exert a degree of control on the physical environment. The development and implementation of models and simulations are very important components of the control process. By collecting and processing real-world data about the world, scientists and technicians create virtual maps, models, and simulations, allowing them to experience the physical environment from within the controlled confines of laboratories. Engagement with models and simulations allows scientists and engineers to perfect techniques like space flight, controlling the flow of a river, or the management of natural resources before experiencing the real thing. Scientists create hundreds of models and run thousands of simulations in an effort to get the desired results in real-world situations (Latour 1987, 238–239).

Indigenous geographic knowledge and histories are often positioned at the periphery. There are several reasons for this marginality. First, indigenous histories are found within oral histories in oral tradition of storytelling. As a result, not too many historians have engaged directly with indigenous storytellers and their knowledge of historical geographies. Second, the geographies, histories, and stories are embedded within the signs and symbols that appear on maps, pictorial ledgers, pictorial calendars, works of art, and even petroglyphs on the side of landforms. As stated above, the predominant discourses on indigenous geographies and histories are associated with a grid. Yet, for those willing to put in the effort and the research, there is a possibility of retrieving historical geographies from science, symbols, and stories. There are important aspects of indigenous geographies and histories that cannot be recreated within a grid epistemology that are themselves more central to understanding indigenous historical geographies.

Geographic knowledge systems such as the stories song dance and ceremony hold vast amounts of knowledge. Linda Tuhiwai Smith argued that, "[t]he ideas of contested stories and multiple discourses are valued ways of knowing within indigenous communities. These contested accounts are stored within genealogies, within the landscape, within the weavings and carvings, even within the personal names that many people carried. The means by which these histories were stored was through their systems of knowledge. Many of these systems have since been reclassified as oral traditions rather than histories" (Smith 1999, 33). Indigenous historical place narratives are "performed and dramatized, [becoming] a form of narrative art, a type of historical theater in which the pastness of the past is summarily stripped away and long-elapsed events are made to unfold as if before one's eyes. It is history given largely in the active present tense" (Basso 1996, 33). In other words, history is simultaneously past and present and indigital HGIS is sociotechnical.

4.5 Conclusion

HGIS are hybrid systems that combine qualitative and quantitative data sources. I refer to hybrid, indigenous HGIS as indigital geographic information networks or the amalgamation of indigenous, scientific, and technological knowledge systems, characterized as fragmentary, contradictory, and full of uncertainties. For academics, working with native languages and geographies is analogous to assembling a puzzle. Geographic knowledge is often difficult to access or recall. For American Indians, total recall of the past is impossible because our past is always hidden from us for reasons including social amnesia or the absence of public images, all containing only fragments of our experiences or "partial recall" (Lippard 1992). Kiowa writer N. Scott Momaday echoes this condition by arguing that "[t]he verbal tradition by which it [history] has been preserved has suffered deterioration in time. What remains is fragmentary: mythology, legend, lore, and hearsay—and of course the idea itself, as crucial and complete as it ever was. That is the miracle" (Momaday 1969, 7).

The ideas and materials making up an indigenous HGIS are not static, pure, and homogeneous entities. Rather, systems are heterogeneous, open, dynamic, and subject to change. Five hundred years of encounters and exchanges contributed to the current condition. Language is one example of an indigenous knowledge system that can transmit geographic knowledge. The current state of orally transmitted languages is precarious. There are efforts within North American Indian communities to revitalize language through oral storytelling activities and phonetic spellings of vocabulary words. However, written forms of language are increasing in usage. Universities and scholarly research are important for building stronger and more durable language and geographic knowledge networks. Kiowa is only one example. Additional research is needed to more fully understand the impacts that university research and curriculum development are having upon the revitalization and practice of native languages in North America. Indigenous communities will have to determine how they will use their languages in the future. However, digital technology offers one trajectory of research and supports multiple applications.

References

Basso K (1996) Wisdom sits in places: landscape and language among the Western Apache. University of New Mexico Press, Albuquerque
Bijker WE, Hughes TP, Pinch T (eds) (1987) The social construction of technological systems. New directions in the sociology and history of technology. MIT, Cambridge
Bray F (1997) Technology and gender: fabrics of power in late imperial China. University of California Press, Berkeley
Brey P (2003) Theorizing modernity and technology. In: Misa TJ, Brey P, Feenberg A (eds) In modernity and technology. MIT, Cambridge, pp 33–72
Cajete G (2000) Native science. Clear Light Publishing, Santa Fe
Callon M (1986) Some elements of a sociology of translation: domestication of the scallops and the fishermen of St Brieuc Bay. In: Law J (ed) Power, action, belief: a new sociology of knowledge? Routledge, London, pp 196–233
Callon M (1991) Techno-economic networks and irreversibility. In: Law J (ed) A sociology of monsters. Routledge, New York, pp 132–164
Craig W, Harris T, Weiner D (eds) (2002) Community participation and geographic information systems. Taylor & Francis, London
Crampton J (2010) Mapping: a critical introduction to cartography and GIS. Wiley-Blackwell, Oxford
Curry M (1998) Digital places: living with geographic information technologies. Routledge, London
Dyson L, Hendriks M, Grant S (eds) (2007) Information technology and indigenous people. Idea Group Publishing, Hershey
Goodchild M (2007) Citizens as sensors: the world of volunteered geography. GeoJournal 69:211–221
Greene C (2009) One hundred summer: a Kiowa calendar record. University of Nebraska Press, Lincoln
Gregory IS, Ell PS (2007) Historical GIS: technologies, methodologies and scholarship. Cambridge studies in historical geography. Cambridge University Press, Cambridge
Harley B (1989) Deconstructing the map. Cartographica 26(2):1–20
Harris T, Weiner D (1998) Empowerment, marginalization and "community-integrated" GIS. Cartogr Geogr Inf Syst 25(2):67–76
Harvey F (2000) The social construction of geographic information systems. Int J Geogr Inf Sci 14(8):711–713
Harvey F (2001) Constructing GIS: actor networks of collaboration. URISA 13(1):29–38
Harvey F, Chrisman N (1998) Boundary objects and the social construction of GIS technology. Environ Plan A 30:1683–1694
Holton G, Berez A, Williams S (2007) Building the Dena'ina language Alaska archive. In: Dyson L, Hendriks M, Grant S (eds) Information technology and indigenous people. Idea Group Publishing, Hershey, pp 205–209
Inglebert E, Banks S, Pavel D, Friedlander R, Stone M (2007) Multimedia curriculum development based on the oral tradition. In: Dyson L, Hendriks M, Grant S (eds) Information technology and indigenous people. Idea Group Publishing, Hershey, pp 123–125
Knowles A (2005) Emerging trend in historical GIS. Hist Geogr 33:7–13
Laituri M (2011) Indigenous peoples' issues and indigenous uses of GIS. In: Nyerges TL, Couclelis H, McMaster R (eds) The SAGE handbook of GIS and society. Sage, London, pp 202–211
Latour B (1987) Science in action: how to follow scientists and engineers through society. Harvard University Press, Cambridge
Latour B (1991) Technology is society made durable. In: Law J (ed) A sociology of monsters. Routledge, New York, pp 103–131
Latour B (2005) Reassembling the social: an introduction to actor-network-theory. Oxford University Press, Oxford

Law J (ed) (1991) A sociology of monsters. Routledge, New York
Law J (1999) After ANT: complexity, naming and topology. In: Law J, Hassard J (eds) Actor network theory and after. Blackwell, Oxford, pp 1–14
Leavy B (2007) Digital songlines: digitizing the arts, culture and heritage. In: Dyson L, Hendriks M, Grant S (eds) Information technology and indigenous people. Idea Group Publishing, Hershey, pp 159–169
Lewis G (1998) Cartographic encounters: perspectives on native American mapmaking and map use. University of Chicago Press, Chicago
Lippard L (1992) Partial recall: photographs of native Americans. New Press, New York
Mark D, Turk A (2003) Landscape categories in Yindjibarndi: ontology, environment, and language. In: Kuhn W, Worboys MF, Timpf S (eds) Spatial information theory: foundations of geographic information science, international conference, COSIT 2003, Ittingen, September 24–28, 2003, Lecture notes in computer science, vol 2825. Springer, Heidelberg, pp 28–45
Martin E (2000) Actor-networks and implementation: examples from conservation GIS in Ecuador. Int J Geogr Inf Sci 14(8):715–738
Momaday N (1969) The way to rainy mountain. University of New Mexico Press, Albuquerque
Murdoch J (1995) Actor-networks and the evolution of economic forms: combining description and explanation in theories of regulation, flexible specialisation, and networks. Environ Plan A 27:731–757
Murdoch J (2006) Poststructuralist geographies: a guide to relational space. SAGE, London
Palmer G (2003) Telling stories the Kiowa way. University of Arizona Press, Tuscon
Palmer M (2009) Engaging with indigital geographic information networks. Futures 41:33–40
Palmer M (2012a) Theorizing indigital geographic information networks. Cartographica 47(2): 80–91
Palmer M (2012b) Cartographic encounters at the BIA GIS center of calculation. Am Indian Cult Res J 36(2):75–102
Palmer M, Hanney J (2010) Geographic information networks in American Indian governments and communities. Int J Virtual Communities Soc Netw 2(2):1–10
Palmer M, Rundstrom R (2013) GIS, Internal Colonialism, and the U.S. Bureau of Indian Affairs. Annals of the Association of American Geographers. In press
Pearce M (2008) Framing the days: place and narrative in cartography. Cartogr Geogr Inf Sci 35(1):17–32
Pearce M, Hermann M (2010) Mapping Champlain's travels: restorative techniques for historical cartography. Cartographica 45(1):33–48
Pearce M, Louis R (2008) Mapping indigenous depth of place. Am Indian Cult Res J 32(3):107–126. special Issue: "Mainstreaming Indigenous Geographies"
Pickles J (1995) Ground truth: the social implications of geographic information systems. Guilford Press, New York
Robins P (2003) Beyond ground truth: GIS and the environmental knowledge of herders, professional foresters, and other traditional communities. Hum Ecol 31(2):233–253
Rundstrom R (1995) GIS, indigenous peoples, and epistemological diversity. Cartogr Geogr Inf Syst 22:45–57
Sappington N (2008) Tribal GIS: supporting native American policies with GIS. ESRI, Redlands
Schuurman N (2000) Trouble in the heartland: GIS and its critics in the 1990s. Prog Hum Geogr 24(4):569–590
Sharp J (2009) Geographies of postcolonialism. SAGE, London
Sheppard E (1995) GIS and society: towards a research agenda. Cartogr Geogr Inf Syst 22(1):5–16
Sheppard E, Couclelis H, Graham S, Harrington JW, Onsrud H (1999) Geographies of the information society. Int J Geogr Inf Sci 13(8):797–823
Sieber R (2006) Public participation geographic information systems: a literature review and framework. Ann Assoc Am Geogr 96(3):491–507
Sieber R, Wellen C (2008) Blending participatory GIS and geo-spatial ontologies for indigenous knowledge preservation. http://www.emse.fr/site/SAGEO2007/CDROM/CQFD14.pdf
Smith L (1999) Decolonizing methodologies: research and indigenous people. Zed Books, London

Soja E (1996) Thirdspace: journey to Los Angeles and other real-and-imagined places. Polity Press, Oxford

Sugito S, Kubota S (2007) Alliance project: digital kinship database and genealogy. In: Dyson L, Hendriks M, Grant S (eds) Information technology and indigenous people. Idea Group Publishing, Hershey, pp 260–265

Warhus M (1997) Another America: native American maps and the history of our land. St. Martin's Press, New York

Winner L (1980) Do artifacts have politics? Daedalus 109:121–136

Wood D (2010) Rethinking the power of maps. Guilford Press, New York

Chapter 5
Place in Time: GIS and the Spatial Imagination in Teaching History

Detlev Mares and Wolfgang Moschek

5.1 Introduction

No matter how space may be experienced in current time, the experience of historical space is always a matter of the individual imagination. If you try to follow the itinerary of a medieval king, you may make good use of maps depicting his favorite residences, you may call upon images of medieval castles in your mind, and you may study the design of medieval means of transportation. And yet, the space you try to imagine is irretrievably past. In your mind, modern images and concepts will be entangled with scraps of information from past ages, and the result will always be a mental representation of former space rather than anything approaching an immediate impression of historical space itself.

However plausible these considerations of imagined spaces may seem, when talking about historical space, we tend to take its existence for granted. Past events are easily placed within a spatial framework of reference that remains blurred and unspecified, as if space was a preexisting entity without its own history and with unchanging features and points of reference. Therefore, the aim in dealing with space in history teaching has to be twofold: Learners have to understand that space has evolved over time, and they have to be made to reflect on the imaginative quality of their own views of past space. Both aims can be achieved by using an HGIS in history teaching, both at primary-secondary school and university levels. GIS provides the opportunity to arrange historical data with spatial references in manifold and ever-changing ways, thus enabling the learner to reconceptualize a

D. Mares (✉)
Technische Universität Darmstadt, Darmstadt, Germany
e-mail: mares@pg.tu-darmstadt.de

W. Moschek
Private Tagesheim- und Internatsschule Dieburg, Dieburg, Germany
e-mail: gis@pg.tu-darmstadt.de

A. von Lünen and C. Travis (eds.), *History and GIS: Epistemologies, Considerations and Reflections*, DOI 10.1007/978-94-007-5009-8_5,

given set of data from different perspectives. In the process, space is experienced in its imaginative qualities and loses its appearance of a preexisting, unquestioningly given entity. The fallacy of the existence of historical space without consideration of the mechanisms of the mental representation of space is exposed, and different images of space become available. This process encourages further questions about the quality of historical space and its changes over time. In this chapter, we show how GIS can be used at different teaching levels to achieve a deeper understanding of historical space and its imagined qualities. First, we will discuss teaching spatial thought and briefly introduce the main features of GIS. Then we will turn to establishing a set of aims for historical spatial learning, before dealing with the use of GIS in history teaching. The conclusion will summarize the benefits to be gained from a stronger focus on a spatial dimension in the teaching of history and the possible uses of GIS in achieving these aims.

5.2 Development of Spatial Thinking and Historical Thinking in Childhood

Space and time are fundamental categories which help shape human understanding of environment. History and space structure human perception of the world and are essential in shaping identities related to one's place in society. Time in particular inscribes a historical mode into the functions of the human mind. The development of spatial and temporal thinking starts in the early stages of childhood and is refined all through an individual's life. Developing spatial imagination in history teaching thus contributes to basic processes of identity building in childhood and provides crucial skills for dealing with the challenges of life.

In this section, we will provide definitions of the terms "spatial thinking" and "temporal" or "historical thinking" and relate both terms to the intellectual development of children. With these definitions the psychological foundations and the mental framework of "learning time and place" by a didactical and methodological approaches can be developed. In the following sections, we combine both aspects within the context of "historical spatial thinking," before we move on to show how GIS can be used as a teaching tool.

5.2.1 *Spatial Thinking*

Spatial thinking constitutes a universal mode of thinking, which can be analyzed from different disciplinary perspectives. "Thinking spatially" requires three elements which include: concepts of space, tools of representation, and a process of reasoning (National Research Council 2006, ix). These three elements have to be integrated into the spatial competence teaching in schools and universities in order to improve an understanding of static, dynamic, and—in our case—spatial historical relationships.

Every human being develops its own geographical sense of space. First, like an island-view, the individual's surroundings are represented by the individual places where life occurs: from own room to house, school, or playground. In later stages, connections between the islands emerge and are reflected as routes, which form the representation of space in later development phases. This sketch conforms to more sophisticated observations on the development of spatial thinking. Piaget and Inhelder divided such a development into three steps: topological, projected, and Euclidian space (Piaget and Inhelder 1999).

1. The topological phase is egocentric and concerned with simple topological relations in words like "divided," "inside," "outside," "open." This phase typically ends at the age of eight.
2. In the projected phase, children start to connect their knowledge of close places to larger spaces: Objects and simple topological relations are incorporated correctly into larger systems of orientation, but straight lines which connect places are the main ordinates of orientation.
3. At about the age of nine, children start to incorporate horizontal and vertical lines into more complex topographical relationships. At about 12 years, the normal child reaches a level of spatial perception comparable to that of adults. Proportions and spatial distances are developed in the same way as the relationship between action, space, and the perception of space.

Piaget's theory on the development of cognitive thinking maintains that between 12 and 15 years of age, children are able to make hypothetical assumptions, and to question and solve these assumptions in logical or illogical form.[1] GIS can help students to improve and develop these respective spatial levels of thinking.

5.2.2 *Historical/Temporal Thinking*

What about the historical component of thinking? How does consciousness of time develop? It seems it develops in similar fashion as spatial awareness does.

Conducting an empirical study, Shemilt (1984, 14) deduced that development of historical thinking progressed through four levels:

1. At level 1, historical narratives are devoid of logic and lack meaning other than that those attached to discrete episodes. "Evidence" and "information" are not distinguished at this level (Schniotalle 2003, 42).
2. At level 2, historical narratives seem to obey a simple iron necessity.

[1] Like Stückrath (1955, 28–50), who tested children and students, cf. Rinschede (2005, 72–76).

3. At level 3, history is construed as complex and thick textured.
4. At level 4, the concept of "historical period" is more fully developed and linked to a narrative logic that is bound to time and place (Shemilt 1987, 46).

As S. Wineburg observes, complex historical thinking is connected with the generation of epistemological beliefs (Wineburg 1991, 511). Parallels between the development of geographical and historical thinking are obvious: Both temporal and spatial dimensions, for example, the teaching of history and geography, mutually reinforce each other if implemented within the context of teaching historical spatial thought. This synergetic effect provides an understanding of what time and space are composed of, which connections they have, and how to navigate both dimensions competently during the course of a lifetime.

History and geography at their different levels of human reception can be easily supported by a GIS. Beginners can map their own home, school, or other important places in a GIS and construct a database with different variables to convey aspects of this limited space, for example, size, inhabitants, pictures, Internet links. The historical aspect could be filled with old pictures of the place, information about former inhabitants, etc. Even during the first steps of learning how to think spatially and historically, GIS with its layer and multimedia-based approach can support students on their respective levels of development. Entries for this approach can be either a spatial or a historical topic. In what respect, then, is historical spatial thinking more than just the sum of its component parts, temporal and spatial thinking?

5.3 What Is Historical Spatial Thinking?

Historical spatial thinking does not deal with time and space in the sense of geological change but in the temporal dimension of the human interaction with space.[2] This includes a twofold perspective: firstly, as the consideration of the effects of human interference upon spatial arrangements and, secondly, as the effects of spatial preconditions on human actions—without claiming the priority of space in a geo-deterministic fashion. Moreover, historical spatial thinking involves an awareness that a view situated from the present allows for different representations of past human action in historical space.

The main obstacle to a clear conception of historical spatial thought is the multifaceted character of space. Spatial arrangements can be discussed with reference to natural space, languages spoken in a given area, economic, social, and political aspects, as well as other themes and subjects. The historian V. Oswalt has suggested a typology of space in historical thinking and offers the following categories (Oswalt 2010, 227–231)[3]:

[2] An essential approach to this topic is Bollnow (2011).

[3] Cf. also Jekel (2006, 26ff).

1. Sectorally structured space which is linked in a defined point of time to a different subject like politics, religion, culture, economy, etc. These subjects are important aspects of our society past and present and thus important in historical thinking.
2. Dimensionally structured space, which handles space and history from micro to macro levels, from simple single rooms to towns and to regional, national, and worldwide scales and areas of interest.
3. Interwoven space, which reflects the importance in the change of perspective, for example, from center to periphery, as well as underlining the rising complexity of the connections between the different sectors and dimensions of space and time in history.
4. The space of historical actions and plots. This refers to historical "events" subordinated to phrases like "first crusade," "civil war," or "crossing the alps with an army." Variables such as time, duration, places, and actors may change, but within this category it becomes very clear that space is not only a container for historical processes but even more an integrating factor for the production of history itself.
5. The space of natural environment. This is a prerequisite to all human actions and developments in space, and can be summed up with a "challenge and response" (Arnold Toynbee), such as reflected in Frederick Jackson Turner's *Frontier-Thesis*, and its effect on US society.
6. The space of mental mapping. This focuses on the individual reconstruction of place in time and space.

As far as teaching historical spatial thought is concerned, the existing literature and curricula in different countries conceptualize such phenomena in different ways and suggest different approaches to teaching. However, without much doubt, the dominant approach consists of the interpretation of printed historical maps in order to develop and evolve historical spatial thinking. In Germany, for example, working with maps and historical atlases is a standard feature of historical teaching textbooks.[4] Other possibilities include the use of historical pictures (paintings, drawings, or photos) or texts/sources, such as itineraries from antiquity to the present.

In terms of methodology, understanding maps is the foundation upon which any in-depth historical spatial thinking has to be built. However, we argue that GIS offers teaching opportunities that go beyond the use of maps. Looking at the various typologies of historical space, GIS supports nearly all of them. While conventional maps are mostly static, GIS is dynamic; while conventional maps have to be analyzed with reference to the aims of their makers, GIS turns the users themselves into the map makers. However, much more importantly is the one crucial feature of GIS technology: It is not—in the first place—simply a device to produce maps but a powerful tool to visualize data. This quality is an essential and intrinsic value for teaching historical spatial thinking.

[4]See Sauer (2010, 234); or Black (1992, 508).

5.4 GIS: A Tool for Spatial Thinking

It has long been observed that GIS offers a useful support tool for the development of spatial thought. Since its applicability is obvious in contexts dealing with geo-referenced data, discussions on the value of GIS in education have centered on the teaching of geographical sciences.[5] In 2006, a report by the National Research Council discussed the potential benefits of GIS for educational purposes in the USA. The authors presented GIS as a tool that can be used in "a variety of educational settings," to be employed "throughout the curriculum or in traditional subject-based curricula" as well as "in all grades" and across a range of teaching modes, such as "individual and stand alone, collaborative, or networked" (National Research Council 2006, 218).

At the same time, much attention was given to the limitations of using GIS in schools. Doubts concerned the infrastructural preconditions, technical support, the need of instruction for teachers and pupils, and a general lack of customization of the device for educational purposes (National Research Council 2006, 218–219). Although several authors have endeavored to allay these worries,[6] they are not without justification and have resulted in calls for "the redesign of GIS" as an "educational software" (National Research Council 2006, 220). However, empirical examples of successful GIS-based teaching units give sufficient proof that challenges connected with the implementation of GIS in education can be and are met in practice.[7] We therefore leave this important discussion to other quarters and focus instead on the more specific question of how to use GIS to support the development of spatial thinking in historical contexts. This requires a brief discussion on the transformation of GIS into a tool for the analysis of historical space, before we move to an outline of the potential of GIS for teaching historical spatial thought.

5.5 GIS in Historical Spatial Thinking: The Making of HGIS

If a "temporal" dimension is added to the spatial core of GIS, the geographical information system can be transformed into a historical geographical information system (HGIS). Pioneers started to envision HGIS methodologies in the 1990s, for example, in the field of urban history (Schott and Jung 1997; Schott 1999), but the theoretical and practical implications of connecting space and time in GIS have only been worked out more recently. The collections edited by Anne Kelly Knowles were milestones in presenting case studies that show the practical viability

[5]Cf. De Lange (2006), Donert (2007) and Green (2001).

[6]See De Lange (2006), Püschel (2007) and Jekel (2006, 31).

[7]See Gebelein (2009) or Brauckmann (2011); for a more general view, see Churchill and Hillier (2008).

of employing HGIS (Knowles 2002, 2008). Substantial projects developing the historical dimension of GIS have emerged, such as the China Historical GIS (Bol 2008) or the website "A Vision of Britain through time,"[8] while Ian Gregory and Paul Ell have explored the uses of a temporalized GIS for historical geography (Gregory and Ell 2007). Areas with marked affinities for the use of HGIS include environmental history, military history, and historical analyses of archaeological evidence. Moreover, the history of religion, migration history, and even time-honored political history have been broached as possible subfields that could profit in the future from increasing applications of HGIS (Bodenhamer 2008, 230).[9]

GIS has not been primarily designed to answer historical questions. So why bother to use a tool whose core structure does not integrate time? More precisely, what is the specific historical surplus value of using GIS? In many historical areas of interest, one may detect a spatial dimension. GIS enables researchers to gather data from diverse historical sources into databases and visualize the data—combined in different ways—in the form of maps (Goodchild 2008). This means the map produced in GIS is not necessarily the final result of the research project, but rather a preliminary tool that raises new questions and calls for further analysis, for example, by new visualizations of further data combinations. Any historical question that relates to space may find a prospective answer by a professional handling of GIS. For example, people long wondered about the purpose of the "Limes," a Roman frontier system in many places of Europe and parts of Northern Africa. The use of GIS allows historians to establish the connections between land use and military security measures that informed the placing of farms, fortresses, and lines of transport (Lünen and Moschek 2011).

While the practicality of GIS for historical research is now beyond doubt and a lively community of "gistorians" has evolved,[10] concepts on the uses of GIS in gistory education, both at school and university level, are much less developed.[11] In this paper, we focus on the benefits GIS offers on two levels of historical learning: Firstly, on a theoretical level, GIS provides fundamental insights into the conceptualization of time and space. As mentioned above, in traditional history education, the spatial dimension usually is represented by maps. Pupils learn to read and interpret maps that have been specially designed to visualize certain circumstances or causal connections from the past. Less frequently, pupils are made to engage with historical maps which focus narrowly on items preserved from the past. Beyond historical geography, the latter technique can be used to uncover world views and the image-making processes of our ancestors. Teaching with HGIS may integrate some methods from traditional map analysis (Böttcher 2002, 2007), but this is not the crucial point. GIS-generated maps represent "intermediate stages"

[8]http://www.visionofbritain.org.uk, last accessed 23 Sep 2011.

[9]See also David Bodenhamer's contribution to this volume.

[10]For "gistory" and "gistorians" see the contribution of von Lünen in this book.

[11]For some recent examples, see Lünen et al. (2011), Churchill and Hillier (2008) and Plöger (1999).

in the context of a more comprehensive investigation (Lünen and Moschek 2011, 249).[12] Since GIS does not "draw" maps but helps to visualize data, the potentially preliminary character of GIS-generated maps in teaching contexts means that a continual process of data re-collation is set in motion. A single map produced in an attempt to reconstruct the situation of the past is interpreted and may be immediately deconstructed in the process, giving way to additional data visualizations. Thus, new maps may answer some of the questions they were created to answer, or these maps may call for yet more re-collations of data. By transforming self-generated maps into heuristic tools in the process of answering questions, pupils come to understand that representations of past space are constructions which depend on the perspective inherent in the choice of data. Thus, pupils get a sense of agency and learn to question their preconceptions about time and space by probing different narratives. This emphasizes the positive dynamics of creating representations of space and time with GIS.

GIS thus makes a valuable contribution to the development of spatial thinking. There is no such thing as "absolute space" (Jekel 2006, 25). V. Oswalt has delineated three attributes in an individual's perception of space ("drei Eigenschaften der Raumwahrnehmung"):

1. As an individual construct;
2. as a multidimensional attribute (e.g., integrating smell or sound reminiscences);
3. as an identity-building attribute (e.g., by separating "our" space from "their" space).[13]

Teaching historical spatial thought has to help students become aware of the preconceived images of historical space which are lodged in their minds (mental maps) and furnish them with the intellectual tools to reflect on spatial representations they encounter in everyday life. GIS can be a crucial teaching instrument for achieving these aims. On a pragmatic level, GIS offers the opportunity to design modules that correspond with many of the general aims of historical teaching, such as problem solving and other various aspects.

5.6 GIS and Teaching the Construction of Historical Space

Teaching with HGIS is more than "time" plus "space" in connecting various chronologies and place gazetteers. A majority of lessons at school or university start with just simple facts about time and place/space. In many cases, students will never attain a level which allows them to conduct critical and constructive reconstructions

[12] See also Alexander von Lünen's contribution in this book for more on this.

[13] Oswalt (2010, 223–225); see also Tuan (2007).

of history and its spaces.[14] Reading spatial descriptions is one of the more advanced goals in history teaching, and together with a GIS additional learning channels can be opened.

As previously mentioned, students' capabilities and their level of spatial and historical thinking have to be evaluated and integrated into the didactics of spatial history education. Depending on the level/age, the scaling, and the extension of space, different levels of argument, sector, dimension, and plot must be chosen. The use of plat or cadastral maps for history teaching can be very useful while teaching micro- or mesohistory. Overlaying maps of the area of a school or part of a hometown can show their geographical and historical development in simple ways. Together with aerial photos or other historical maps of the same areas, students can reconstruct, by using different map layers, the spatial elements which constitute a history. The Geoportal of the French *Institute Geographique National* is an impressive example for this type of database and sources for the study of French spatial history.[15] Another example is the David Rumsey Map Collection, which allows the transfer of historical maps to Google Earth.[16] In other countries, this kind of access is rare and sometimes expensive. But even the simple possibility of combining time and space with changing maps (cities, landscapes, countries etc.) is more than a simple printed historical map or atlas can offer. Even by using these GIS-based Internet platforms, young students can playfully change the appearance of space by choosing different layers and thus experience how familiar places have changed over time. Overlay techniques can help students to understand what lies temporally "under" or "over" a certain place. To combine data from an old map with additional historic content and historic cartographic information can equally show contrasting views and representations of space. One example is the transformation of the content of the ancient fourth century Tabula Peutingeriana (c. 675 cm long and 34 cm in height) to Google Maps as a route-planer.[17] The confluence of these two spatial visualizations into one map image enables even pupils with little experience of spatial historical thinking to recognize the variations between different historical levels of spatial representation. The possibilities of Google Earth and other similar platforms are also limited. But due to their availability and "ease of use," they offer a very good start. Web-GIS applications induce a discovery-based learning process and can inspire curiosity in students for more information. For historical spatial thinking, these online platforms fascinate by the amount of data they provide and the extension of possibilities. One disadvantage is that pupils often get lost somewhere in the virtual world; so a strict frame of questions, layers, and data can guide younger pupils, who are not experienced in handling larger spatial data and

[14] For a more detailed analysis of the concepts of "reconstruction" and "deconstruction" in history, see Munslow (1997).

[15] http://www.geoportail.fr/, last accessed 23 Sept 2011.

[16] http://www.davidrumsey.com/, last accessed 23 Sept 2011. Or a fairly new site, "Old Maps Online": http://www.oldmapsonline.org, last accessed 07 Mar 2012.

[17] http://www.omnesviae.org, last accessed 23 Sept 2011.

complex historical connections, to focus on a sensible aim. A "closed" scalable desktop GIS with connections to different online-data sources is one solution.

From a basic level, where all data and questions are provided for students, up to the most advanced levels where students frame their own questions, define the space of analysis, and collect data, the aim of the HGIS should be kept in mind. The idea of using GIS in history is to provide an alternative tool for students to analyze and reconstruct historical spaces to further historical learning. By using GIS in this manner, discovery-based learning is incorporated.

5.7 Discover!

When pupils begin to use GIS to investigate historical events, they can use its various tools out of the box to visualize their subject. Either doing this with or without GIS software technical experience or with a strong-guided frame of questions to address, pupils learn to produce different results by using various GIS tools. By using the same data and the same historical event/space, they can come to understand that the result of their GIS construction depends upon an individual view and use of methods. GIS allows students to change color, scale, and amount or range of data in a fast and easy way. With a few "clicks" students can discover a variety of historical spatial reconstructions. In addition, users can easily discover the different modes of expression inherent in different forms of visualization,[18] from the choice of colors (e.g., using red to emphasize) to the size of symbols up to the manipulation of given data and the display of statistical analysis (Monmonier 1991). All these factors can be very useful not only for a student's reflection on historic spatial events but also for basic presentation skills.

5.8 Troubleshooting?

The method of problem solving-based learning (finding the problem, building a hypothesis, developing solving strategies, problem solving, verification—falsification or modification of the hypothesis—reflection)[19] is well recognized as a very sustainable method in teaching. The basic skills of analysis and reconstruction in geography and history are nevertheless essential for this method. The scalable level of effort employed as a methodology in history, geography, and GIS supports pupils

[18]For a discussion of the term "visualization" in historical scholarship cf. Staley (2003), see also the interview with David Staley in this volume.

[19]Rinschede (2005, 62–64), De Lange (2006, 12), Zumbach and Jekel (2006, 35) and Lünen et al. (2011, 62).

in reaching the aims of the lesson or the project.[20] Even in one lesson, with given data and questions, pupils are able to recombine, color, depict, or describe their *own*, self-generated historical maps. While the pupils themselves create maps for a given problem, they have to decide which data they want to use to solve the problem. By doing this, the pupils build—deliberately or not—a hypothesis and think about the choice of data or their presentation to solve the problem. With the advantage of a computerized visualization through GIS, it is easy to use, reuse, and change the results of data recombination on a historical-geographical reconstruction by means of a few "clicks." In this manner, pupils acquire the competence of dealing with the multiple dimensions involved in visualizing historic spaces much better than with any well-produced static map. More than this, with the digital output made of re-arrangeable data—which distinguishes a GIS from a consecutive text—HGIS is nonlinear and multidimensional, allowing pupils to jump between different graphics, texts, and pictures. In this context, the dynamic use of data reinforces the skills of problem solving-based learning and promotes and fosters analytical thinking.[21]

5.9 Making Maps: Constructing Space

One more benefit: At the end of a lesson, pupils get a solid result by means of a saved/printed self-produced map which can further inspire their interest in learning. HGIS is a result-oriented teaching tool: Even no result is a result because the reason for this "failure" can form part of a reflection on the method, the procedure, the data used or the main question entertained. Moreover, specific skills in the presentation, reception, and construction of historic space are trained, analyzed, and visualized (Oswalt 2010, 231). In comparison to the normal interpretation of a historic space by a historical atlas or a wall map, the contrast becomes obvious. With a HGIS, an increased level of mapping competence and critical questioning on historical topics is possible. Of course, normal teaching methods, that is, reflecting on the hermeneutics and heuristic parsing of historical sources, should not be neglected—these methods remain fundamental.

5.10 Thinking of HGIS: Thinking Global?

Therefore, working with HGIS can induce modern teaching methods or at the least enhance more traditional kinds of teaching. GIS brings new approaches to the teaching of history and geography, making both subjects more interesting, more modern, and even more available and consumable (Taylor and Duran 2006,

---

[20] Schäfer (2006, 60) and Püschel (2007).

[21] Churchill and Hillier (2008, 86ff) and Ott and Swiaczny (2001).

11ff). This availability has another beneficial value: Autonomous learning can be promoted through access to a global network. Working with history and space in a HGIS can provide a new platform of tools which connects traditional methods of history and historical reconstruction, to a constructive multimedia application which can give pupils an understanding of how a modern globalized and Internet-linked world functions.

5.11 Some Conclusions: The Historical Imagination of Space and Teaching with HGIS

The importance of possessing map skills in history cannot be doubted. Despite the fact that the traditional use of historical maps and the skills is still very useful, with a HGIS, the interactive interpretation of historical maps can be employed by learners at any age (Horstkemper and Sorbello Staub 2010). The pedagogical benefits of a constructive, easy to edit, multimedia, and multidimensional HGIS have been mentioned. Didactical planning within history and geography curricula should emphasize and integrate the interdisciplinary foci of a HGIS.

More challenging is the practical setting for GIS in history lessons or research. The technical side is becoming less difficult. Not only better graphic and user interfaces, faster computers and Internet connections make using GIS easier. Also, very powerful GIS freeware is on the market (such as QuantumGIS[22]). The main problem remains that the practical education of teachers in the use of GIS is almost non-existent, even at university level. Sophisticated projects on university level exist but they focus on high-level research (National Research Council 2006, 220). Examples include using GIS to investigate the Tabula Peutingeriana (Talbert and Elliot 2008) or the transformation of Ptolemaeus' list of places and distances from his ancient world depictions into modern maps (Kleineberg et al. 2011). Though conducted in a very scientific manner, such projects translate historical human reception and construction from ancient times to our contemporary spatial perspective through the reconstructive methods of GIS.

In the last 10–15 years, historical GIS has been the topic of a rising number of discussions in pedagogy, didactics, and methodology, and has produced many fruitful but singular projects. To apply Piaget's levels of skill of spatial reception in the discipline of history, we are still functioning between levels 2 and 3. A few more important steps need to be implemented before a general reception, construction, and practice of HGIS is more effective and widespread in teaching history.

[22] http://www.qgis.org

References

Black JM (1992) The historical atlas: teaching tool or coffee-table book. Hist Teach 25(4):489–512
Bodenhamer DJ (2008) History and GIS: implications for the discipline. In: Knowles AK (ed) Placing history: how maps, spatial data, and GIS are changing historical scholarship. ESRI, Redlands, pp 219–233
Bol PK (2008) Creating a GIS for the history of China. In: Knowles AK (ed) Placing history: how maps, spatial data, and GIS are changing historical scholarship. ESRI, Redlands, pp 27–59
Bollnow FO (2011) Human space (trans: Shuttleworth C). Hyphen, London
Böttcher C (2002) Die Karte. In: Pandel HJ, Schneider G (eds) Handbuch Medien im Geschichtsunterricht. Wochenschau-Verlag, Schwalbach, pp 170–196
Böttcher C (2007) Umgang mit Karten. In: Mayer U, Adamski P (eds) Handbuch Methoden im Geschichtsunterricht. Wochenschau-Verlag, Schwalbach, pp 225–254
Brauckmann S (2011) Kulturlandschaftsforschung im Geschichtsunterricht? Raumzeitliche Orientierung durch GIS-gestützte Nahraumanalyse. Zeitschrift für Geschichtsdidaktik 10:66–77
Churchill R, Hillier A (2008) Teaching with GIS. In: Knowles AK (ed) Placing history: how maps, spatial data, and GIS are changing historical scholarship. ESRI, Redlands, pp 61–94
De Lange N (2006) Geoinformationssysteme in Schulen – derzeitiger Stand und zukünftiger Einsatz. In: Jekel T, Koller A, Strobl J (eds) Lernen mit Geoinformationen. Herbert Wichmann Verlag, Heidelberg, pp 11–22
Donert K (2007) Geoinformation in Education – A European perspective. In: Jekel T, Koller A, Strobl J (eds) Lernen mit Geoinformationen II. Herbert Wichmann Verlag, Heidelberg, pp 3–19
Gebelein J (2009) Blending history with geographic information system technology: using USS Arizona GIS data to engage students in technology and history of the Pearl Harbor attack. Geogr Compass 3:2011–2024
Goodchild MF (2008) Combining space and time: new potential for temporal GIS. In: Knowles AK (ed) Placing history: how maps, spatial data, and GIS are changing historical scholarship. ESRI, Redlands, pp 179–197
Green DR (ed) (2001) GIS: A sourcebook for schools. Taylor & Francis, London
Gregory IS, Ell PS (2007) Historical GIS: technologies, methodologies and scholarship. Cambridge studies in historical geography, Cambridge University Press, Cambridge
Horstkemper G, Sorbello Staub A (2010) Klio im Kartenraum. Historisch-geographische Informationssysteme und digitale Geschichtskarten. Geschichte in Wissenschaft und Unterricht 61:259–260
Jekel T (2006) Virtuelle Flüge, räumliches Problemlösen, kritisch-konstruktive Didaktik – Anforderungen an Lehren und Lernen mit Geoinformation. In: Jekel T, Koller A, Strobl J (eds) Lernen mit Geoinformationen. Herbert Wichmann Verlag, Heidelberg, pp 23–34
Kleineberg A, Marx C, Knobloch E, Lelgermann D (2011) Germania und die Insel Thule: Die Entschlüsselung von Ptolemaios' 'Atlas der Oikumene', 2nd edn. Wissenschaftliche Buchgesellschaft, Darmstadt
Knowles AK (ed) (2002) Past time, past place. GIS for history. ESRI, Redlands
Knowles AK (ed) (2008) Placing history: how maps, spatial data, and GIS are changing historical scholarship. ESRI, Redlands
Lünen Av, Moschek W (2011) Without limits: Ancient history and GIS. In: Dear M, Ketchum J, Luria S, Richardson D (eds) Geohumanities: art, history, text at the edge of place, Routledge, London, pp 241–250
Lünen Av, Mares D, Moschek W (2011) Im Raume analysieren wir die Zeit. Geo-Informationssysteme in der historischen Forschung und Lehre. Zeitschrift für Geschichtsdidaktik 10:56–65
Monmonier M (1991) How to lie with maps. University of Chicago Press, Chicago & London
Munslow A (1997) Deconstructing history. Routledge, London
National Research Council (2006) Learning to think spatially. GIS as a Support system in the K-12 curriculum. The National Academies Press, Washington, DC

Oswalt V (2010) Das Wo zum Was und Wann. Der spatial turn und seine Bedeutung für die Geschichtsdidaktik. Geschichte in Wissenschaft und Unterricht 61:220–233
Ott T, Swiaczny F (2001) Time-integrative geographic information systems: management and analysis of spatio-temporal data. Springer, Berlin/Heidelberg
Piaget J, Inhelder B (1999) Die Entwicklung des räumlichen Denkens beim Kinde. In: Gesammelte Werke, vol 6, 3rd edn, Klett-Cotta, Stuttgart, p 25ff.
Plöger R (1999) Anwendung Geographischer Informationssysteme (GIS) für historisch-geographische Aufgabenstellungen. In: Ebeling D (ed) Historisch-thematische Kartographie: Konzepte, Methoden, Anwendungen. Verlag für Regionalgeschichte, Bielefeld, pp 9–23
Püschel L (2007) Vom Web-GIS zum Desktop-GIS – ein didaktisch-methodisches Konzept zur Einführung von Geographischen Informationssystemen an Schulen. In: Jekel T, Koller A, Strobl J (eds) Lernen mit Geoinformationen II. Herbert Wichmann Verlag, Heidelberg, pp 138–148
Rinschede G (2005) Geographiedidaktik, 2nd edn. Schöningh, Paderborn
Sauer M (2010) Zur 'Kartenkompetenz' von Schülern. Ergebnisse einer empirischen Untersuchung. Geschichte in Wissenschaft und Unterricht 61:234–248
Schäfer D (2006) Skalierbarer Einsatz von Geographischen Informationssystemen (GIS) in Schulen. In: Jekel T, Koller A, Strobl J (eds) Lernen mit Geoinformationen. Herbert Wichmann Verlag, Heidelberg, pp 60–69
Schniotalle M (2003) Räumliche Schülervorstellungen von Europa. Tenea, Berlin
Schott D (1999) HIST – Entwicklung und Anwendung eines historischen Geo-Informationssystems für die Stadtforschung. In: Ebeling D (ed) Historisch-thematische Kartographie: Konzepte, Methoden, Anwendungen. Verlag für Regionalgeschichte, Bielefeld, pp 38–47
Schott D, Jung V (1997) HIST: Ein Geo-Informationssystem für Stadtgeschichte. In: Arbeitsgemeinschaft Geschichte und EDV (ed) Geschichte und EDV: Probleme und Fortschritte – Probleme mit dem Fortschritt?, Gawl, Bochum, pp 15–28
Shemilt D (1984) The devil's locomotive. Hist Theory 22(4):1–18
Shemilt D (1987) Adolescent ideas about evidence and methodology in history. In: Portal C (ed) The history curriculum for teachers, Falmer, London, pp 39–61
Staley DJ (2003) Computers, visualization, and history. How new technology will transform our understanding of the past. Sharpe, London
Stückrath F (1955) Kind und Raum – Psychologische Voraussetzungen der Raumlehre in der Volksschule – Psychologie der Unterrichtsfächer der Volksschule. Kösel, München
Talbert RJA, Elliot T (2008) New window on the Peutinger map of the Roman world. In: Knowles AK (ed) Placing history: how maps, spatial data, and GIS are changing historical scholarship. ESRI, Redlands, pp 199–218
Taylor JA, Duran M (2006) Teaching social studies with technology. New research on collaborative approaches. Hist Teacher 40(1):9–25
Tuan YF (2007) Space and place: The perspective of experience, 5th edn. Minneapolis University Press, Minneapolis
Wineburg SS (1991) On the reading of historical texts: notes on the branch between school and academy. American Educational Research Journal 28(3):495–519
Zumbach J, Jekel T (2006) Problemorientiertes Lernen mit Geoinformation – Ansätze und Beispiele zum erfahrungsbasierten Lernen. In: Jekel T, Koller A, Strobl J (eds) Lernen mit Geoinformationen. Herbert Wichmann Verlag, Heidelberg, pp 35–46

Chapter 6
"Thou Shalt Make No Graven Maps!": An Interview with Gunnar Olsson

Alexander von Lünen and Gunnar Olsson

It is not a bad definition of man to describe him as a tool-making animal.
(Charles Babbage: The Exposition of 1851, 1851)

Gunnar Olsson (b. 1935) is an acclaimed human geographer from Uppsala, Sweden. He received his first degree in 1960 and his doctoral degree in 1968, both from the University of Uppsala. He was a Visiting Fellow of St. John's College, Oxford; fellow of the American Council of Learned Societies at the University of Pennsylvania; and Assistant Professor at the University of Michigan, before becoming Collegiate Professor at the same institution; in 1977 he returned to Sweden, working for Nordplan, a "consortium university" run by the Nordic countries. In 1997 he became Professor at the University of Uppsala.

His main academic concern was planning and its consequences. Starting from a strong analytical and quantitative approach, his initial research topic was the relation between "distance" and human interaction. He soon came to realize that this would not suffice to grasp the underlying methodological and moral questions. Attending a lecture on econometrics by Swedish economist Hermann O. Wold in 1962, Olsson would learn that even the simplest statistical concepts—such as correlation or linear regression—carry with them intrinsic philosophical problems about epistemology and causality. He soon got caught up in the question whether spatial variation was cause or effect of social interaction. While reasoning about this issue, Olsson came to realize that questions in geography such as these give way to deeper inquiries into form and process, sign and signified. Working with stochastic models in his doctoral thesis, he pursued to tackle the "geographical inference problem", that is,

A. von Lünen (✉)
Geography Department, University of Portsmouth, Portsmouth, UK

G. Olsson
Kulturgeografiska Institutionen, Uppsala Universiteit, Upsalla, Sweden

A. von Lünen and C. Travis (eds.), *History and GIS: Epistemologies, Considerations and Reflections*, DOI 10.1007/978-94-007-5009-8_6,

what inferences we can draw from geographical information. A few months after he defended his thesis, however, Olsson became suspicious about such enterprises, realizing "that even the most prefect description of a spatial pattern cannot be used as the basis for inferences about how and why it came about" (Olsson 2002, 252).

These issues, the relation between form and process, began to occupy his thoughts and led to a first monograph being published in 1975: *Birds in Eggs*. It dealt with both epistemological and ontological issues in geographic reasoning—a topic so novel that the Library of Congress shelved it under "ornithology" (Olsson 2002, 256). He soon ventured further into issues of language, philosophy, semiotics and culture, using images from mythology to explain modern power relations. The investigations in linguistics and philosophy in the 1970s were soon superseded with linguistic experiments in his work in the 1980s. These linguistic investigations were first and foremost conducted to find a way of conveying the invisible in maps; his linguistic experiments, on the other hand, were not supposed to be an artistic enterprise but rather the desire to create a geographic language which in turn would create a novel approach to social science (Philo 1984, 218).

Olsson would soon become one of geography's most acclaimed thinkers, ranking level with eminent scholars such as Yi-Fu Tuan, Anne Buttimer or Nigel Thrift (cf. Abrahamsson and Gren 2012). His quest was to write what might have been Immanuel Kant's *Fourth Critique*, a critique of cartographical reasoning, to which end *Abysmal* was published in 2007. Olsson, albeit not an expert in the field, also looked into GIS and remote sensing and their value for this kind of reasoning. While acknowledging these technologies as important tools, he criticizes its contemporary use: "The only crux is that the majority of the GIS techniques draw on observed autocorrelations, not on insight into causal relations" (Olsson 2002, 253). Rather, he encourages GIScientists, should GIS and remote sensing be seen as artistic tools, in which the colour patterns created by them should be analysed in the same way paintings are, "For what is remote sensing, if not a human activity located in the interface between poetry and painting? What is a satellite picture, if not a constellation of signs waiting to be transformed into meaning-filled symbols?" (Olsson 1999, 141).

Alexander von Lünen (AvL) met Gunnar Olsson (GO) at his home in Uppsala, Sweden, on 24 November 2012 to speak with him about the insights we can gain from maps and GIS—in general, and more specifically in history.[1]

6.1 Maps and Art, Images and Imagination

After we had some introductory conversation, we somehow arrived at the topic of art, something that Gunnar Olsson is vitally interested in.

GO: What connection do you see between your particular interest in GIS and creating art?

[1] The interview was held in English and edited for publication.

AvL: I haven't thought about it too deeply. But on the other hand, arts and maps have something in common. As in art the story behind that map is often more interesting than the map itself; the process how the map was created and what the map was meant to represent, rather than the actual representation which resulted in that process. It is the same with paintings. Sometimes you look at a painting and it is not very interesting, not very appealing—visually! But when you learn the story behind it, you think: o yeah, that is quite interesting. I mean, that's what you deal with in a lot of modern art. You mentioned the Ready-Mades in your recent book[2]; where you look at the thing and think: not very interesting, it's just an every-day item, but then you learn about the idea behind it, that concept of Ready-Made Art; and you think: that's quite intriguing, actually. So you start reflecting about the things that it tries to tell you, rather than just looking at the object.

GO: This raises a very, very interesting issue: can you have imagination without image? The natural opinion of most people is that you can. But only some people say there are physicists, for instance, who can do their research without constructing images. So, this again becomes some sort of ontological issue, because even when we think of the most abstract concepts we think about them as if they were objects. And when you do that we commit thorough ontological violence.

So what is imagination? It is the ability to make present what is absent, and vice versa. For physics it is of course what a physicist does: it is to make present of what is absent and vice versa. So when we perform these tricks, which are of course some magic trick in a sense, it is very similar to what I called "cartographic reasoning".

AvL: Yes, I see your point about art and cartographical reasoning. Art makes us think, makes us reflect. Or as Susan Sontag once put it: "Real art has the capacity to make us nervous."

GO: And it is fantastically interesting when it comes to the map now. That in the second and third commandment you have a double prohibition: a prohibition against graven images, which is about pictures, and you have a prohibition against using the ruler's name in vain. It is a prohibition against telling the true story of God. So here you see that in these commandments, you can read that as a complete censorship; censorship against the understanding of God. And now, the interesting thing is that a map is at the same time as much a story as it is a picture. Now you understand why maps are so blasphemous. Why you end up in Siberia, why you are killed, because you have a particular type.

---

[2]"Ready-Mades" refers to an art form where everyday objects are presented as objects of art. Sometimes the objects are artistically enhanced, either by painting on them or by unusual combinations of objects, for example, but by and large they remain untouched. The point being that the observer is made to think about everyday items—and therefore everyday life. Probably the best known Ready-Made artist was Frenchman Marcel Duchamp (1887–1968). Gunnar Olsson refers to Ready-Mades to point out that we never start from scratch in our imagination but rely on established signs (Olsson 2007, 158).

Our moral understanding is much based in invisible maps. And they imagine this world. So, once that is said, you're not in the world, but it brings us straight into the whole issue of imagination. You cannot conceive a space as such; it is natural that space is somehow "pictured".

AvL: That is often the way historians use maps: treating it as some kind of "text" that conveys a story from the past. I would not identify a GIS with a printed map per se, though. Because with a GIS you create the maps yourself, and probably just for yourself. So it is a bit self-referential, if you like. Whereas with a historic map you always have to look at the context in which the map was created, both in terms of the technology and the intent—the whole setup. So we differentiate between GIS and maps in that respect. And because it is inevitably interdisciplinary to do historical GIS, history and geography at least involved, I am very much interested in—both in terms of this book, but also in my general work—how geography and history inform each other, and what is necessary to benefit from one another. Is it really just a "translation problem" between the two subjects, or is it more fundamental?

GO: It is more fundamental. I think that this noise is due to rhetoric and GIS. And the approach, of different yet related subjects that interact with each other. And they do! And then you have to split it up, in rhetorical terms, in metaphor and metonymy, and then you see how these two are related. It will seem that geography ought to be richer in metaphors, and history richer in metonymy. But the question is: are they equal or are they not?

And, of course, GIS is in a sense a combination of pictures. And then it becomes very dangerous that these GIS statements go hand in hand with cartographic subjects. At the same time, you have a new way of understanding.

AvL: Yes. What I find so intriguing about GIS for historians is that ... I mean, historians have worked with maps for a long time; you know, maps as a source like a written document, a written statement. So they took maps from other people to work with, but GIS actually gives one the tool to create your own maps. So, rather than being consumers of maps, they become producers of maps. This gives them a way to actually reason about their subject in a wholly different way than they could do before. And that's what intrigued me as a historian, to be able to basically play around with the spatial configuration.

The question is: how far can you leave the sources behind and venture into the realm of imagination? That goes a bit into the direction of what Carlo Ginzburg published in a book, where he looked at the "fact" and "reality" in history.[3] Because the postmodern historians say: there are no facts in history, it's all a narrative. So there is no truth in history, all the accounts are embedded in a certain ideology. And Ginzburg makes that reference to Aristotle, saying: yes, it's chiefly rhetoric, but a good rhetoric is based on facts, and you can't make any sense out of facts without rhetoric. So the two are twins.

[3]Cf. Ginzburg (1999).

GO: They are twins! We cannot make up just anything! I think you can make a few stories up, but you cannot liberate the stories from the object, from the material, that it is subconscious, that it becomes just fantasy. Even when you map the invisible, like idols or belief systems.

6.2 Of Models and Representations

AvL: But then you open up a whole new can of worms: historical narratives. When you translate them into a computer model, you get a "lost in translation" issue again, literally. So how representative is what you get from a GIS for historical research?

GO: If you take the word "representation" ... literate, it's a fantastic image, because it has two prefixes: it is both "re" and "pre". Both looking back and looking forward. And that is exactly what "representation" is. So therefore, it is not the "re", not the "pre" that is important, but the crucial thing is in the "sentation" part. Because the prefixes can only do what prefixes do: they fix you in a sentence. So, here you have the fixed points—"pre" and "re"—already in the representation, which I think is why also what I call the cartographic reason is so important. Because you cannot have cartographic reason without fixed points and rotations.

Now, if you take the medieval maps, the *Mappae Mundi*; then the fixed point on those maps will be the Garden of Eden. Why so? Because the fixed point is nowadays of course the magnetic North Pole. But it didn't used to be that way, therefore in the *Mappae Mundi* what is "up" on the map is the East, and everyone wanted to go and find the Garden of Eden, of course. And since then it shifted around, but it was always in the East. And that is why we have the term "orient" for finding the way. Because you orient yourself to the Orient, to the East. So, now, when the signification/reification has proceeded, then we should really not be talking about orienting ourselves in the world, we should say "northing" ourselves. Why do I mention this? Because this shift in fixed-points between the medieval maps and later maps illustrates how we hang the world on a hook. Telling us, and perhaps not even themselves, what it is. To get, in a sense, into people's mind-set with cartographic thinking.

Here it's much of an issue of "boundary" and "imagination". We know, or we think that we can approach a limit only from the inside. But since we are imaginative and impulsive, we think that this limit has two sides. And even though we realize we cannot reach to the other side, we cannot stop imagining what might be there. And that in the Roman case here, and in the Greek case, was of course the emptiness it solved. That is where language did not reach, where understanding didn't reach, and that is where the tension of course is. So, these edges are very, very imaginative. If you now come to the mathematical geography, especially Spatial Analysis, it is fantastic that you have a new area in which you have a limit: it stops with your data, hence at the edge. It creates a lot of problems.

AvL: It does indeed! And I take your earlier point about "imagination" and "translation". To abstract into a data model, we have to both translate and imagine. Abstraction involves a loss of information; it is an aggregation, so we have to imagine what information might be useful and which one is not. It is quite difficult for us to imagine and abstract at the same time. That goes a little bit in the direction what David Hume said, in *Treatise on Human Nature*: even though we talk in abstract terms we have a concrete image in our mind. Let's just say we say the word "horse", which is very abstract, because that doesn't refer to any particular breed or colour. But when we say the word "horse" we still have in our mind a concrete horse, which we saw the other day maybe. That's what Hume said: that we use concrete terms as if they were abstract when they aren't.[4] And that's the same in geography, and it's definitely the case in history, when people try to apply abstract concepts to historical events and actors, who had concrete concepts in mind.

GO: And in the Renaissance, when we were rediscovering and reworking and reformalizing the perspective. And there is a difference, a distinction between "perspectiva naturalis", which is in the vanishing point and the "perspectiva artificialis" being the point of meaning, which is the artificialis: I *see* what you mean. So in some ways your problem, in my understanding, is the real challenge we are facing—and have always been facing, but we are facing more now than ever before—is our inability to be abstract. To sort of loosen this modal painting that ties us to the physical image, but this ties us to the eye senses. And therefore one issue that lies at the heart of whatever I am doing is the relation between the five senses of the body and the sixth sense.

So in a sense, it's the inference problem in another way. And it is the relation between materiality and realizing that to communicate the meaning we have to thingify. A very close relation between reification and deification. Which again raises the issue of how we accumulate knowledge.

6.3 Do We Think for Our Tools or Do Our Tools Think for Us?

AvL: Isn't the equipment what determines how we think about the world, or our subject, rather than these fundamental questions? If you look for instance at scientists: science has changed so dramatically when laboratories were introduced in the nineteenth century. It changed the whole way of thinking

[4] "Abstract ideas are therefore in themselves individual, however they may become general in their representation. The image in the mind is only that of a particular object, tho' the application of it in our reasoning be the same, as if it were universal" (Hume 1964, 327–328).

and how they work, how they go about their subject. And I argued in a paper a couple of years ago that this introduces also a generational issue.[5]

In the nineteenth century there was a big debate between scientists, called the "Mechanism-Vitalism debate": whether you can actually discover the human nature or the organism simply by going into the laboratory or if you have to consider philosophical or theological issues as well. That was what the vitalists said, the "older" professors if you will, that you can't understand the human body without the "vital force" which is behind the chemical and physical processes in the body. Whereas younger ones who "grew up" in the laboratories said: no, all we need to worry about is how we get the laboratory to a level where we can understand the human body better, where we learn more about the biology, i.e. the chemistry and the physics of the body, rather than the supposed spiritual force behind it. And I argued that it wasn't really an ideological battle, it was just that the younger physicists, the younger biologists, grew up in the laboratory so that they knew nothing else than this form of research. So they didn't worry about these spiritual issues.

And I wonder whether it is not the case in geography as well. You now have people who grow up with computers and GIS and Remote Sensing, it determines their way of thinking about geography, doesn't it?

OG: Very, very much so. They are lacking a kind of history that you automatically think, of course, that everyone before us were like us. We see ourselves as the lab dog used in testing, even if we know that the dog is not like us. We cannot understand the dog if we don't think it is like us.

AvL: That has a lot in common with historians' practise: how much empathy do we need to have towards the historic actors we investigate? As historians we have the blessing of the hindsight, yet we have to consider the contemporary notions historic actors had, i.e. how much they knew as opposed what we now know about them.

And then, obviously, we encounter the limit not only of our thinking, but also of our tools. What could a GIS do for us once we squeezed the data into it? Is it supposed to deliver answers or should it just assist us finding them?

GO: The tools order us what to do. It is not we who rule the tools, but you end up often with the tools ruling us. And, logic, of course is one prime example. And then there is a temptation to overuse the tools, to (over)extend them. So you have to know very well when to stop. The trick is not to stop, but to know when to stop.

AvL: I think that's the problem with some historical GIS people: They don't know when to stop. It is just from beginning to end all about the GIS. They use it as a kind of black box; you know, they put the data in and you get a result and the result is the final one. Whereas I argue that the result you get from GIS is an intermediate step, you still have to make sense of it. It can assist in what you do, but it is not giving us the answer to our problem.

[5]Cf. Lünen (2006).

GO: But here again you have … you are back in imagination. Because in order to stop in time you have to have the ability to imagine what has not yet arrived. There is an "already", but not yet. That is the art to *see*: to see the whole image that has not yet been delivered.

AvL: Well, in economics, in game theory, there is the concept called the "multi-armed bandit". You know, the one-armed bandit, the gambling machine; and then imagine you can have a machine—or rather: several machines put together—so you have N arms, an arbitrary number of arms (each representing a variable in a complex system, such as the economy). And the idea with each machine is that it follows a pattern. So if you observe it long enough you will find that pattern and so you know the strategy to play that machine and to win money; and the question, the problem—the "multi-armed bandit problem"—is: how long do you observe it? Because every time you observe it, you have to play it, i.e. you have to put in some money (and spend some time). So how much money are you going to spend before you can actually start exploiting your observation? And is it then still worth playing the machine, because you have to put so much money in it to gain the pattern, to observe the pattern, that you spent so much money already, you never going to get it out again by playing it. Where do you draw the line? Where do you stop observing and where are you going to start exploiting your observations?[6]

GO: If you begin this debate you have to look at what information is: it is forming your image. That is what information is. It is when one uses metaphors. Metaphors are images but then become information. Here is a problem with translation, because we cannot have metaphors without metonymy. So metonymy is in some sense the dynamite, and the metaphor is the charge. Metaphor is essentially a picture, or an image, that starts in your imagination. And that it's impossible to stop. But here we can have a metaphor, say, in English, and the first two steps in this association chain can be quite similar, but then it diverts. That is why translating poetry is so incredibly difficult, because poetry uses so many metaphors. In translating the original poem, you might actually need a quite different metaphor for the poem to make sense. So, coming back to the representational issues that we were talking about earlier …. So in one setting we use this term X, and in another setting we use a boundary stone (i.e. description of what it isn't, rather than what it is), to get to the same type, to get up to human relations. Which are the problems when we have archives. In archives you have a number of potential metaphors.

AvL: And then there's still the question of different people using the archive. That's why all … that's the main difference between historians and geographers—the debate about history and geography, and the relation between the two—, that's obviously because they have different traditions.

[6]See Berry and Fristedt (1985) for a discussion of the "multi-armed bandit problem".

Whereas I say: the difference in tradition is not about the way of thinking, but more about the way of reading. They don't think differently, they read differently.

GO: Historical geography in the classical sense is, of course, about "reading the landscape". Reading the traces in the land. And there, Remote Sensing can be a very, very useful tool. So you have two ways of understanding, of representing the world: representation by pictures, or through words.

AvL: Indeed! And geographers do the former, whereas historians are into the latter. In historical GIS, it seems, it is also representing through numbers. Which has severe limits, obviously.

6.4 Considerations

GO: It is why I changed course in my approach. That is why in the 1960s I went the way I did: it was the edge of that approach I encountered.

If you look at this "quantitative revolution" in the 1960s, that was clearly the root for this dissatisfaction. It was prevalent in my generation. No science, no rules discovered, nothing. So this was some sort of physics, with models which came from physics, but the use was of course social engineering. This idea of calibrating parameters of the social model, so that you could then know how to change society to the better. And we used very simple linear regression. And then, of course, you can measure all sorts of things: x went down, the slope went up, etc. So the focus was on studying that slope. It was stable, but then it turned out it wasn't. It varied with time and space. Economically, socially ... in there, there were very high correlations. Some confuse that observation with prediction. This parameter was clearly made to the hierarchy of central places. So, obviously, migration to and from big cities were less distance sensitive.

So then it turns out that these were not ways of capturing and describing social interaction, but the spatial distribution of it. So it was not that we could map life in a prism, but we mapped the prism itself. We started with deterministic models ... but then re-formulated it in to a stochastic model. And it was with these stochastic models that then one could find very good ... could translate spatial distributions and discontinuities into quite precise stochastic models. And then the question was: how did that distribution arise? For that we used different models. And it was through that medium when this moment had first been translated into a stochastic model, and it was possible to "see" the distance problem. So you see here, it was an attempt to put into the model a knob for individual variation. That was not to aggregate, but to differentiate. So it was, at least in my case, it was stripping "aggregate prediction" down to individual disaggregated explanations. Thus getting to the cause, not to the result. Not to the aggregate result. So, you may get very high correlations, but not so high explanatory power. That is alright if you stick to descriptions, but if you move from descriptions to prescription then this is different. Because if you have prescription without having your theory based in causation, then you are bound to fail.

So now I have dug a little deep into these issues that have—not in my case, actually—got into the abyss. It is a long history, not so complicated, but it has clear steps; where I realized that it will be impossible to understand the inference problem with knowing these roots. On different levels, but not to any deep level. And of course "causation" is "story". And it is much, much later that I got into these Greek and Classical topics.

AvL: I see your point. On the other hand, I am sometimes having this rather Buddhist stance of, you know, I don't worry about causation, I take the world as it is. But, obviously, you get to the point where you think: Now I have to dig deeper to see actual ....

GO: Especially, as GIS gets so deeply involved in practical action. That is why I bring this up, because it is important not only for understanding change. It is important to foresee the consequences of the actions that have been prompted by the GIS. You have to re-read or foresee the consequences, particularly the unintended consequences. And this is why I and my colleagues in the generation before me ... Marx said: the philosophers try to understand the world, but it is important to change it. It is therefore important to understand the world and then—perhaps—to change it.

But I also realize that every tool has its limits. And one can overuse it. So, a laser beam can be a fantastic tool, but it can also mean death.

AvL: Obviously, a tool is never "neutral". I mean, it was built with a purpose, so it determines the work we do with it. Every tool was built with a purpose, and if you look at the history of a GIS, it was built with a purpose in mind: environmental planning and social planning. So even though you can appropriate that tool to a different purpose, it still has that heritage, that initial aim in it ... people forget about the "background story" of a tool.

GO: Which is not "quod erat demonstrandum", but "quod erad absiendis".

AvL: That's the bottom line, or, the premise, or the 101 of the history of science and technology. Technology is a social product—I think Karl Marx said that.

GO: Very much he said that.

AvL: Yes, and that's why a lot of historical GIS people ... Humphrey Southall had that idea with the Great Britain Historical GIS, which kind of defected from the mainstream GIS approach, and rather works as a digital gazetteer.[7] And in the beginning a lot of people were saying: oh, that's not actually a GIS, because it doesn't have these and these features. But then we argued: well, that's only the features you are used to, because ESRI has put them out there for you. Because they have the main market share in GIS products, they determine what a GIS is, and what it isn't. Whereas if you look

[7] Cf. Southall et al. (2011).

into a computer science dictionary, GIS is actually a very broad and basic definition: it is any information system that has ....

GO: It is so broad that it is hard to say for me what it is. You know what it is, but for me it is very vague.

AvL: It is one company that defines what GIS is and that's what most people know—both geographers and non-geographers—and take for granted. GIS has to have these kind of features, or it's not a GIS. And that is a very big obstacle for the debate in historical GIS, I mean, most people in historical GIS realized: ok, we have to abandon the ESRI model, but what do we do as an alternative? So, still, when they discuss alternatives, it is the ESRI data model, the ESRI paradigm that overshadows the debate completely. The fixed point, if you will.

6.5 Reflections

AvL: This brings us back to our point of departure: representation. What do you think would be necessary to actually use maps in a better way? Do we have to re-wire our minds, or use different tools? Because when you say, we have to go to the bottom of the problem ... and then what?

GO: Sometimes I think that the same reasoning what I have been able to do with cartographic reasoning ... it may be, but it is a reasoning that is already dead. So it is clear that these philosophers ... Kant in particular.... When he wrote his three critiques, he himself of course used it for a formal reason, that in my book was cartographic reasoning ... but this was so close to him he could not see what the reader could ... that he could write three critiques but not the fourth. If it is now beginning to be possible to see at least the need for the fourth critique, then I would say that it is because it has lost its real power.

It is agreed that it is not very powerful or very interesting, but it is not so much at the edge that it is really dangerous for the world itself, or our understanding. The world has changed so much that cartographic reasoning for that new world is obsolete. It doesn't matter if you have, say, quantum physics, that does not mean that Newtonian physics is irrelevant to the world. Of course it means that the cutting edge issues are different. I have a feeling that we have something that is the same situation in the social sciences. And that therefore cartographic reasoning in this conventional way is a way of capturing, if you will, the "Newtonian" social sciences. Not quantum physics. But then, what is coming instead? Hard to say, except that what comes instead is almost certain to build on what we already have. So quantum physics isn't a complete rejection of everything else, of course. Then you should look for changes, transformations—whatever it is in these ... and that is exactly what I am doing. And then the question is which of these three is the most important one? That's to say: what is the hypothesis to get into first? That are the fixed points. And the most difficult position in between is the translation process, the scale.

I am trying to see what this cartographic reasoning, which is so fantastic, has served us so well. You ask: how can I experiment? How can I transform that fantastic mode of inferences so it can help us also? I have a feeling that perhaps this is helping to locate what the problem is and to locate where the problem is, then we can search for ways to deal with it. Then you go on from there.

You see how deeply difficult it is, because it has to do with the socialization processes whereby we take for granted this form and procedure—it's in the educational system. So it is a pedagogic enterprise. Take Plato's *Republic*, for example. What is the point of the *Republic*? What is it trying to do? He is ... it is a pedagogic tasks that he has set himself. Because he says: what is a good life in a good city? And how do we educate a good citizen for the good city? So it's a pedagogic tract, set up for how to organize this in teaching. He had the same problem, in a sense.

Now if you specify and present this current situation in these terms, then the question is: what does it mean? How does one handle this? Or how is the world handling this? This is nothing you and I can do, this is much deeper. And then one runs the risk that whatever we do, we rather cement the current road blocks.

AvL: Indeed. The issue is not the absence of the right tools, but the absence of creative thinking about the tools. For instance, we did a ... a colleague and I, when I was still in Darmstadt in Germany; that colleague did ancient history, the Romans in the Germanic territories and how they set up their settlements and how they colonized the Germanic territories. And we used a GIS to literally get a different view on that story.[8]

We put archaeological data into a GIS and mapped it—you know, showing the border walls they erected. And our premise was that from that spatial configuration you can actually infer the mind-set of the people who erected those buildings, simply by the way they place buildings in a strategic way ... you know, the walls, the border walls they set up, which was completely new at the time. At this time they didn't have clear-cut borders, you usually used rivers or hills to delineate your territory. And also there was no "kingdom" of Germany at the time, so they didn't face a uniform opponent. There were scattered tribes, who were raiding the Roman territory, but they didn't wage a full-scale war against the Romans. But nonetheless, the Romans did erect those walls, and there was no big force on the other side they would have to keep away. And that was our point of departure: to use a GIS to look at this kind of setup and to infer what the Romans had in mind when they erected those walls. And it worked quite well for our case. Basically, we

[8] See Lünen and Moschek (2011) and the chapter by Detlev Mares and Wolfgang Moschek in this book. Also von Lünen's chapter.

found out that the walls weren't a military building; they were more like a cultural building. It was more like that the Romans put it up for themselves, to tell themselves: that's where our world ends, and after that there's the barbarians. So they didn't build the walls to keep the tribes on the other side, but to show to themselves where their turf ends.

And then we actually went so far to produce maps deliberately that don't represent reality—we used geometric analysis techniques, and the maps didn't represent reality at all, but they were just an aid if you want, a bit like a notepad for us to reason about the cultural and spiritual thinking of the Romans. Because we didn't have any written source that would have said "We built the walls, because of ...". So we took it from the geographical setup and that worked quite well for us.

GO: That story is a fantastic illustration of what I take to be at the very heart of cartographic reasoning: to start from the (re)presentation of spatial artefacts and from there you infer something about the minds of people.

AvL: So, analytic tools like GIS and Remote Sensing can be used in history, it just needs to be employed differently. But it still begs the question about its general usability. In our case it worked well because the research question—borders—was so topically close to typical uses of GIS.

Other topics may be harder to tackle. It runs the risk of designing the research project around the tool. And as we concluded above: our tools determine our work. And when we speak about "Remote Sensing", that conjures up that old metaphor of technology being an extension to human senses.[9]

We can't help it, but a lot of our imagination and our reasoning is determined by our physicality. That brings us back to the point: how do we abstract? And what do we have to experience first-hand, say, by way of exploration? I was quite baffled when I learned that Immanuel Kant wrote books about Africa, yet he never left Königsberg.

GO: And he got it all wrong!

(laughter)

AvL: Yes, so you're wondering: how can you gain knowledge of things without encountering them? Because, again, that might be a major difference between history and geography. Remote Sensing or not, but at the end of the day you can still go out into the real world and touch the thing you saw on your satellite image—in principle. Something we can't do in history.

GO: Kant clearly took what he imagined and moved it into the "island of truth", where he talked about it as if he was there. So he "misplaced" the truth, actually. We are placing things in the wrong spot. The real question is: whether this has any bearing on Kant's cartographic project, his legacy; that is, that you need to work like a land surveyor to gain knowledge about the

[9]Probably the first one to use this metaphor—technology as extension of human organs—was German philosopher of technology and geographer Ernst Kapp (1808–1898); cf. Kapp (1877).

world. That is a serious question. It is one that he placed in the wrong place. It's sad, but it is not critical. And I have not been able to find out what it is.

He never left Königsberg, but he never was in Königsberg anyway, but everywhere in his mind.

(laughter)

AvL: I am also very interested in the connection between "how much can we learn out of experience and how much abstraction there is". There is another side of the story: you can't just judge things from a subjective perspective, and you always wonder where the middle ground is. As you said about Kant: we learn some things remotely, and some you experience yourself. We have the same problem in history, where we have that problem—predominantly in Contemporary History—where you interview eye witnesses of historic events, something that is very popular in German history: to interview World War Two participants to learn about the history. And that is always a delicate question, because these accounts aren't "neutral". Obviously you remember things differently, or you try to "gloss over" events in the war .... So a lot of historians question whether eye witness accounts are helpful, or if they are not. For instance, the main argument by contemporaries of war atrocities—when you confront them with it—is that they say: oh, you weren't there, you can't judge!

GO: Which is true.

AvL: In one way, yes; in another way you can better judge because you weren't there. So you have these two tensions ....

GO: You can better judge on the level of abstraction, but not on that particular problem.

AvL: Well, it reminds me what Heinrich Heine—a German nineteenth century writer—wrote. He wrote a piece on the history and philosophy of religion in Germany. And in one chapter he deals with a German philosopher named Johann Gottlieb Fichte (1762–1814) and his subjective philosophy: the world as such arises through self-consciousness, i.e. we have to experience subjectively to make statements about the world. And Heine mocks that, saying: well, that's like a monkey cooking his own tail to subjectively cook, rather than to just objectively cook.[10] And that's just what we have discussed about Kant and writing about Africa without having been there. Can you judge better from the outside or the inside?

GO: When put like this, I always come back to solipsism. That is the idea that we cannot know anything about the external world, since we have to observe it through our own sense and can thus make no objective statement about it. The world therefore cannot be scientifically, independently studied.

[10] "This operation reminds us of the monkey seated on the hearth before a copper kettle cooking its own tail; for it is of opinion that the true art of cookery consists not merely in the objective act of cooking, but also in the subjective consciousness of the process of cooking." (Heine 1986, 124)

And if you read the history and development of those advances—on the various issues and questions we have discussed, and to which different authors have different answers—you have a fantastic history. Not only fantastic stories, but also fantastic history. And then you *see* what changes in history ....

References

Abrahamsson C, Gren M (eds) (2012) GO: on the geographies of Gunnar Olsson. Ashgate Gower, Oxford

Berry DA, Fristedt B (1985) Bandit problems: sequential allocation of experiments. Chapman and Hall, London

Ginzburg C (1999) History, rhetoric, and proof. University Press of New England, Hanover

Heine H (1986) Religion and philosophy in Germany: a fragment. State University of New York Press, New York. http://books.google.com/books?id=NnBwQQYywZgC

Hume D (1964) The Philosophical works. A treatise on human nature (vol I: Of the Understanding). Scientia, Aalen. Reprint of the new edition London 1886

Kapp E (1877) Grundlinien einer Philosophie der Technik. Zur Entstehungsgeschichte der Cultur aus neuen Gesichtspunkten. Westermann, Braunschweig

Lünen Av (2006) Who needs scientific instruments? Philosophers! physiology and philosophy in the fin de siècle. In: Grob B, Hooijmaijers H (eds) Who needs scientific instruments. Conference on scientific instruments and their users. 20–22 October, 2005, Museum Boerhaave, Leiden, pp 29–36

Lünen Av, Moschek W (2011) Without limits: ancient history and GIS. In: Dear M, Ketchum J, Luria S, Richardson D (eds) Geohumanities: art, history, text at the edge of place, Routledge, London, pp 241–250

Olsson G (1999) Constellations. Sist Terra 8(1–3):140–144

Olsson G (2002) Glimpses. In: Gould P, Pitts FR (eds) Geographical voices: fourteen autobiographical essays. Syracuse University Press, Syracuse, pp 237–268

Olsson G (2007) Abysmal: a critique of cartographic reason. University of Chicago Press, Chicago

Philo C (1984) Reflections on Gunnar Olsson's contribution to the discourse of contemporary human geography. Environ Plan D Soc Space 2:217–240

Southall H, Mostern R, Berman ML (2011) On historical gazetteers. Int J Humanit Art Comput 5(2):127–145

Chapter 7
Vernacular GIS: Mapping Early Modern Geography and Socioeconomics

Alexi Baker

7.1 Introduction

One of the key objections made against the use of Geographic Information Systems in the study of history is that they cannot contend with the variable and incomplete nature of sources and with the imprecision of maps and other geographical records from the early modern period.[1] Other criticisms are of the cost, complexity, and man-hours which can be involved in setting up the maps and databases necessary to use GIS.[2] As I have discovered during my own mapping of the scientific instrument trade of early eighteenth-century London, these issues can be sidestepped simply by creating digital maps which reflect contemporaries' "vernacular" concepts of physical and socioeconomic space rather than attempting a modern degree of geographical precision: Locations are depicted as contemporaries typically described and understood them, with respect to basic geography and natural and man-made landmarks, rather than striving for precision. In georeferencing (i.e., relating information to geographic location), this is akin to the "informal" descriptions of geographical locations which Linda Hill describes as those "which we use in ordinary discourse using placenames," as compared to "formal" representations which are "based on longitude and latitude coordinates and other spatial referencing systems, which we use in activities such as map making and navigating" today (Hill 2006, 2).

For example, a building in an early modern population center may be described and depicted as existing on a certain street near a certain church or inn, rather than

[1]Gregory and Ell (2007, 1), Corrigan (2010, 76), Knowles and Hillier (2008, 3) and Pickles (1999).

[2]Gregory and Ell (2007, 17, 41, 89), Martí-Henneberg (2011, 11–12), Knowles (2000) and Siebert (2000).

A. Baker (✉)
Department of History and Philosophy of Science, University of Cambridge, Cambridge, UK
e-mail: A_S_Baker@msn.com

A. von Lünen and C. Travis (eds.), *History and GIS: Epistemologies, Considerations and Reflections*, DOI 10.1007/978-94-007-5009-8_7,

in terms of Cartesian coordinates or in terms of the number of meters which lay between it and nearby features. Maps created and interpreted in such vernacular terms can be suitably flexible in the face of records which are uneven or incomplete, if one is researching a moderately large cohort of individuals or entities. They can also create more accurate representations of the ways in which the inhabitants of early modern communities actually viewed space and tried to depict it on paper, which were far more visual and socioeconomically relevant to the inhabitants' experience of place than cartographically precise. This was partially because of the comparatively small size of the spaces occupied by early modern settlements and even by the great metropolis of London. It was easy to navigate the entirety by foot and by other common conveyances, and the buildings and landmarks along many streets could be sighted once one arrived in the general area.

Also during this period, "precision" in all areas of human knowledge and practice beyond just geography and map making was conceived of differently and more varyingly than it is today. It was only as the 1700s progressed that Europeans would make great advances in settling locations and distances on land and eventually at sea through astronomy-based surveying and the so-called search for the longitude—improving maps and charts as a result—and that nations would begin to develop standards of weights and measures. Finally, the programs with which one can create vernacular maps are often made freely or affordably available today through institutions or through the Internet and can be employed by one scholar or student rather than demanding the talents and input of a research team.

Vernacular maps, like the more precise maps which can also be created with GIS, offer a range of benefits for the early modern historian (Knowles and Hillier 2008, xiii). As Ian Gregory and Paul Ell succinctly state with respect to GIS in general: "It provides a toolkit that enables the historian to structure, integrate, manipulate, analyze and display data in ways that are either completely new, or are made significantly easier" (Gregory and Ell 2007, 1). At the center of these advances is the ability of digital mapping to root visual representations of populations of individuals or of socioeconomic entities such as businesses in time as well as location and on both the smaller and larger scales depending upon the nature of the available data and maps. This firstly taps into the so-called "spatial turn" which has, quite rightfully, taken place in history within the past few decades (White 2008, ix). Santa Arias and Barney Warf summarize the point of this development as, "Geography matters, not for the simplistic and overly used reason that everything happens in space, but because *where* things happen is critical to knowing *how* and *why* they happen" (Arias and Warf 2009, 1–2). Vernacular and precision GIS mapping also feed into the recognition that historical analyses mean little without an understanding of change over time. Historical data, place, and time should be viewed as an integrated whole in order to fully understand past and thus present dynamics, as has been emphasized by diverse authors including Langton, Massey, Sack, and Gregory and Ell.[3]

[3]Gregory and Ell (2007, 7, 18–19, 21, 90), Langton (1972), Massey (1999), Massey (2005), Sack (1974), Sack (1972), Bunge and Sack (1973) and Butlin (1993, 51–61).

In terms of the setting of my own research into the production and sale of scientific instruments, the geography, economy, and society of early modern London were inextricably bound together. Communities were sometimes molded by the physical landscape with, for example, the eastern reaches of the metropolis becoming a center for shipping and related crafts and trades in part because ships with masts could not pass London Bridge. However, the characteristics of the socioeconomic groups that settled in different neighborhoods also strongly influenced the physical features there. For example, the trend toward "politeness" among the middling and upper classes during this period influenced the layout and appearance of new streets and buildings in the west and in parts of the central metropolis (Stobart et al. 2007, 3–6). The geography of London was therefore not just its physical form but also the socioeconomic influences and relationships embodied in this landscape. Recognition of this has increasingly molded research in recent decades, such as by informing examinations of the spaces in which shopping for wares such as instruments took place and of the effects of these activities upon the urban landscape.[4] As Arias and Warf suggest, the "spatial turn" encourages such interdisciplinary efforts because "so many lines of thought converge on the topic of spatiality" (Arias and Warf 2009, 1–2).

Digital maps facilitate this by not just pinpointing the locations of people or entities on a representation of the Earth's surface but by also allowing the researcher to attach other kinds of information to these geographical pinpoints. In my case, this included the dates between which members of the scientific instrument trade were at these locations, their livery company affiliation, the types of instruments which they made or sold, and so on.[5] Thus, instead of just being a traditional static map of location, a GIS map is an interactive visual representation of a database in which a researcher can store as much information as they want about their subjects in different "layers." As Gregory and Ell explain, "a layer in GIS is analogous to a table in a database, it is the basic unit of storage for information on one particular subject. It consists of both spatial and attribute data combined [and] usually, in GIS each layer is stored in a separate file" (Gregory and Ell 2007, 36). This means that a vernacular or more precise digital map can be used to easily view parts of a population or of a landscape during specific periods of time, such as one decade or year, and according to different attributes—for example, in my case, all of the scientific instrument makers who belonged to the Clockmakers' Company or who sold globes during a certain decade. The authors go on to emphasize that "as soon as the GIS database is created it can be mapped. This means that the spatial patterns within the data can be repeatedly re-explored throughout the research process, greatly enhancing our ability to explore and understand spatial patterns" (Gregory and Ell 2007, 10–11). This is true of vernacular maps of early modern populations as much as it is of the precision constructs to which the quote refers.

[4]Cf. Stobart et al. (2007), Walsh (1995), Walsh (1995), Davis (1966) and Jeffreys (1954).

[5]The livery companies were trade associations that were given royal permission to regulate their respective crafts in the metropolis, mainly within the original medieval walls, including the Mercers, Grocers, Merchant Taylors and Spectaclemakers.

Depending upon the quality of the map used as the basis for this effort, one can also zoom in or out of the landscape and as easily examine one neighborhood or street as one could an entire city or region. This goes a step beyond the superb parish-wide and regional work already being done in Britain with the aid of basic computer mapping by historians such as Craig Spence (2000a; 2000b). All of this can greatly facilitate the study of populations and locations over time and within their geographical and socioeconomic contexts, as well as the comparison of different populations and locations which have been given similar treatment (Gregory and Ell 2007, 203). This is as true for widespread populations, and for those characterized by socioeconomic rather than by geographic attributes, as it is for geographically constrained communities like those in a city. An impressive example of this is the work done by the recent 4-year research project "The French Book Trade in Enlightenment Europe," overseen at the University of Leeds by Professor Simon Burrows and Dr. Mark Curran. The project used digital mapping and database technology to great effect to map and to analyze the European trade of the late eighteenth-century Swiss publishing house Société Typographique de Neuchâtel.[6] The resulting resources will be made publicly available through the Internet in 2012, allowing anyone to generate new maps and other visual representations of the data.

As in the above case, vernacular and precision GIS can produce clear and engaging resources which can be shared online, in print, or during presentations (Martí-Henneberg 2011, 8). These include maps (and movies or slide shows thereof) of geographical change and of the different socioeconomic attributes of a population. As Richard White says of the purely didactic benefits of using such maps to express dynamics: "Relationships which jump out when presented in a spatial format such as a map tend to clog a narrative, choking its arteries, until—even if the narrative does not expire—the reader, overwhelmed by detail, is ready to die of tedium and confusion" (White 2008, x). As we shall see, these superb visual and analytical tools do not have to be very difficult or expensive to construct, since one can often opt for creating digital maps without recourse to geospatial precision. Especially when studying the early modern era, which was barely affected by such modern concepts, there is no need for such exactness.

7.2 Perceptions of Space and of Socioeconomic Networks in Early Modern Cities

Precision georeferencing has become incorporated into modern life in a wide variety of ways. As Hill writes, these include Internet mapping and GPS guidance, precise and sometimes interactive maps in news reports and print and online publications,

[6]The publications stemming from this project currently include Curran (2010a) and Curran (2010b).

and the increasing commercial and institutional use of GIS to "hold and analyze georeferenced data, leading to the discovery of geographic distribution patterns that support decision makers and planning" (Hill 2006, 2). Of course, vernacular geopositioning persists as well. People continue to view locales in terms of the roles which individuals and individual institutions and places play and have played in their lives, and in terms of visual landmarks which can be noted when giving directions to others, such as a specific shop or a garish sign. However, we must remember that this was the *predominant* way in which urban geography was understood in early modern Europe, before the introduction of civic innovations such as house numbering and before maps were made more physically accurate. There is thus little need for the geographical precision which is often the bedrock of GIS in order to examine populations whose own understanding and representation of space was far more visual and socioeconomic than cartographic.

In London before the second half of the eighteenth century, locations were still conceived of and described entirely in terms of streets, nearby physical and socioeconomic landmarks, and well-known shop signs and proprietors. For example, in 1717, Mrs. Hannam advertised Italian hair dye at the sign of the Three Angels near the Half Moon Tavern in Cheapside.[7] Three years later, George Markham advertised the medicament "the Imperial Essence" at the sign of the Seven Stars under St. Dunstan's Church in Fleet Street.[8] In 1721, Benjamin Workman advertised scientific instruments and medical equipment next to Tom's Coffee House in Russell Street in Covent Garden.[9] Proprietors might also describe their locations in terms of the socioeconomic networks to which they belonged, such as that of a specific trade. These networks were deeply embedded in the physical landscape of the metropolis, with the geographical clustering of occupations and of immigrant or religious groups being quite common.[10] This is partially why digital maps, whether vernacular or more precise in nature, can be such illuminating tools for the analysis and understanding of historical socioeconomic relationships as well as geography. For example, in the case of retail and wholesale shopkeepers, one can use the maps to help to identify: individuals' physical proximity to each other and to other relevant trades, institutions, and communities; the types of passing custom to which they were exposed; and the counterparts with whom they were in direct competition.

An example of socioeconomic relationships and concerns shaping early modern georeferencing comes from the advertisements published by two competing partnerships of opticians in early eighteenth-century London—George Willdey

[7] *Daily Courant* (London, England), 16 February 1717, Issue 4782.

[8] *Daily Post* (London, England), 23 January 1720, Issue 97.

[9] *Daily Courant* (London, England), 28 October 1721, Issue 6247.

[10] See, for example, Gwynn's description of the geographical clustering of Huguenot immigrants in London and Raven's description of the clustering of the London book and print trades near St. Paul's Churchyard. As Power points out, clustering was also influenced by the differing public exposure, rent levels, and types of accommodations to be found not only in different neighborhoods but also in different types of through streets and dead ends. Gwynn (1985, 35, 38), Raven (2004) and Power (1986, 212).

and Timothy Brandreth and their former livery company masters John Yarwell and Ralph Sterrop. Yarwell and Sterrop described their shop at the sign of the Archimedes and Three Spectacles as "the first Shop of that Trade from Ludgate," while the young upstarts Willdey and Brandreth described their shop at the sign of the Archimedes and Globe in Ludgate Street as "the Second Spectacle-Shop above that which was Mr. Yarwell's."[11] Both partnerships described their locations with respect to the local geography of the optical instrument trade, while Willdey and Brandreth also used their location to remind readers of their connections with the well-known Yarwell. The younger men's having opened a shop near their former masters, and having chosen a similar shop sign (i.e., pictorial symbol), suggests that they were trying to directly compete with the older craftsmen.

As previously mentioned, the stage upon which such episodes of London life played out was actually quite a small one, despite the continuing expansion of the capital's population during the early modern period. London was the most populous metropolis in Western Europe by the beginning of the eighteenth century with perhaps 575,000 inhabitants and the most populous in all of Europe by the middle of the century after an increase of at least 100,000 people. Perhaps one in ten English and Welsh men lived in London, as compared to one in forty Frenchmen living in Paris, and as many as one in six English adults lived in the capital at some point in their lives.[12] However, the early modern metropolis only occupied a few square miles of land, even if its borders were always expanding in all directions and on both sides of the Thames. It consisted of "the City," or the central square mile mainly contained within the walls and gates of the original medieval metropolis, and the areas of Westminster, Middlesex, and Surrey that adjoined it.

As a result of such large numbers of people living in such a small area, the total population density of London during this period was about three times greater than that of its inner boroughs at the end of the twentieth century (Guillery 2004, 7). This means that the activities, perceptions, and representations of the capital were molded by its being much smaller (albeit more crowded) than modern cities and being so easily traversed—even if contemporaries thought it gargantuan and labyrinthine.[13] This was essentially true of all of the metropolises and large cities of Europe at this time.[14] In London, inhabitants could easily move about on foot and by carriage, hackney cab, sedans, or waterman. There are differing thoughts on the comparative degree to which members of the different genders and socioeconomic groups

---

[11]*Daily Courant* (London, England), 6 May 1707, Issue 1630.

[12]Beier and Finlay (1986, 1), Wrigley (1987, 134), Rudé (1971, ix, 98), Barnett (1936, 18) and George (1925, 24, 329–330).

[13]Guillery (2004, 7), Merritt (2002, 9), Guillery (2004, 7) and Merritt (2002, 9).

[14]At the beginning of the 1700s, only London, Paris, Amsterdam, Naples, Palermo, Venice, Rome, and Lisbon had more than about 100,000 inhabitants. Vienna, Berlin, and perhaps Lyon reached that point by mid-century (Wrigley 1987, 134).

circulated about the metropolis. However, it is clear that many Londoners from all classes traveled relatively widely for jobs, commerce, religion, socialization, shopping, entertainment, and other interests.[15]

This can be seen in the financial and personal records of innumerable members of at least the upper and middling classes. The diary of the redoubtable Samuel Pepys records his constantly crisscrossing London by foot, coach, and water in pursuit of business and recreation during the second half of the eighteenth century. For example, on September 20, 1668, Pepys walked to church and home, took a coach to visit a beautiful female friend, went to a different church in the hope of seeing a potential paramour, walked to and about Gray's Inn without finding any company, and then ended the day by walking through the fields to Clerkenwell (north of the central city) in the hope of seeing another beauty, before walking home (Pepys 2006, 163). The widespread movement of different types of people for commerce and shopping or leisure is also reflected in the extant shop accounts from early modern retailers and craftsmen. For example, the London-based customers, agents, subcontractors, and business associates listed in the accounts of the aforementioned optician and toyman (i.e., seller of small expensive luxury wares for adults) George Willdey of St. Paul's Churchyard from 1710 to 1725, hailed from almost all parts of the sprawling metropolis north of the river and not just from the central city or the fashionable western neighborhoods.[16]

Thus, London and the other large cities of early modern Europe were mostly perceived visually and at ground and river level and in terms of the socioeconomic networks embedded in the human landscape—not in terms of the clean-cut and precise bird's-eye renderings of a modern map. As Max Byrd said of Daniel Defoe's descriptions of the capital, and Miles Ogborn quoted when discussing cartography, the metropolis in later maps "emerges as curiously featureless, as a collection of names—of streets, buildings, squares—, but not as a realised picture. [...] in the end London exists for us only as a network of traffic, a gigantic system for comings and goings."[17] In applying GIS technology to the analysis of the world before such maps existed, there is seldom a need to attempt geographical precision. Inhabitants of early modern cities lived in a world so much smaller than their modern counterparts that it could be almost entirely walked on foot. Directions more specific than a shop sign and street or proximity to a public landmark seldom needed to be given, since anyone arriving at said street or landmark would likely be able to see their destination or could easily ask for a bit of guidance. Hence, a useful digital map for this time period only needs to pinpoint locations to the street level and only occasionally to building level, and not down to meters of accuracy.

---

[15] Berry (2002, 380–381), (Stobart et al. 2007, 142, 151–152, 155), (Merritt 2002, 10, 145, 149, 152–157, 159).

[16] *Fenhoulet v. Willdey, 1744*, The National Archives at Kew, C 104/21.

[17] Byrd (1978, 12–13) and Ogborn (1998, 31).

7.3 Mapping the Scientific Instrument Trade of Early Modern London

This is clear in my own use of digital maps to depict and to help analyze the scientific instrument trade of London during the first half of the 1700s, for the doctorate which I completed at the University of Oxford in 2010 (cf. Baker 2006, 2009a,b). Although today the products of this eighteenth-century trade are often called "scientific" instruments as a form of shorthand, they were not actually labeled so until well into the next century, after the words "science" and "scientist" adopted their modern meanings and replaced natural and experimental philosophy in learned usage. Early modern instruments were instead classed as optical, mathematical, or philosophical or were specified as being for use in individual subjects such as astronomy and natural philosophy or in "practical" mathematics-oriented pursuits including surveying and navigation (Warner 1990; Field 1988). Most mathematical instruments such as drawing tools, quadrants, and octants had a graduated scale for performing calculations or for measuring angles and distances. Optical instruments employed lenses or mirrors and included microscopes, telescopes, eyeglasses, and some instrument sights. Philosophical instruments were used in the demonstration or investigation of natural phenomena including magnetism, electricity, and the attributes of air. They could be employed in exciting public demonstrations, with electrical machines sparking and sizzling, or air pumps threatening to suffocate small birds and mammals. Many trade members sold multiple types of instruments—as you can see in the trade card in Fig. 7.1 which

Fig. 7.1 A trade card advertising the wares sold by the mathematical instrument and globe maker Nathaniel Hill (working 1746–1764) and likely before him by the mathematical instrument maker and engraver John Coggs (working 1718–1740), both of Fleet Street in London (With permission from the Museum of the History of Science at Oxford)

was likely used by John Coggs of Fleet Street (working in at least 1718–1740) and then by his commercial successor Nathaniel Hill (working 1746–died 1768)—or instruments alongside other wares and services.

During the first half of the eighteenth century, the London instrument trade became the most extensive and renowned instrument trade in Europe, which it would remain for at least a century and a half (Morrison-Low 2007). It was still in its traditional craft-based form, centering upon shops and workshops mainly owned by members of the London livery companies and staffed by apprentices, journeymen, and employees—with a significant contribution of work, materials, and goods coming from outside subcontractors. Most of the surviving information about the trade regards its few hundred known shop-owning members and sometimes their on-site apprentices and employees, which lent itself to my employing geographically oriented technologies from other fields to aid in comprehending and analyzing this population. The same would be true of most other skilled crafts in early modern London, such as clock or jewelry making, and of retail specialties such as the luxury or print trades. Ironically, technologies such as GIS mapping have been very rarely applied to the historical study of technology, despite their success in other fields. One exception is Richard Kremer's mapping of the smaller scale instrument trade which existed along the New England frontier (Kremer 2007).

When I decided to embark upon digital mapping, I consulted Nigel James at the Map Room of the Bodleian Library at the University of Oxford, which serves scholars and students of geography but also other academics interested in incorporating maps and GIS into their research. We decided to employ MapInfo to create my interactive digital maps, which I knew would not require a very high level of geographical precision in order to be productive. In fact, it would have been impossible to pinpoint the locations of most instrument makers and sellers with a high degree of precision, given the aforementioned ways in which the inhabitants of the early modern metropolis viewed and described locations, the changes which have since been made to the streets and buildings of the capital, and the variable nature of the evidence for the specific houses in which trade members lived and worked from sources such as the rate books. I chose to map my research subjects on the well-known map *Plan of the Cities of London and Westminster and Borough of Southwark*, completed by the French Huguenot surveyor and cartographer John Rocque in 1746 and published the following year (Hyde 1981, vii). This was perhaps the earliest map of London to move toward the tidy, geometric, overhead presentation of streets, fields, important buildings, and other landmarks which we associate with modern maps. This did make it easier to assign members to the streets, allies, courts, and other spaces in which they conducted business. The map was in fact unusually precise for its day, with Rocque and the engraver John Pine having employed a surveying instrument known as a theodolite—purchased from the respected instrument maker Jonathan Sisson (working 1722–died 1747) of the Strand—and having garnered public testimonials about the rigor and usefulness of their work from Fellows of the Royal Society (Hyde 1981, v–vi).

We were able to obtain a high-resolution scan of the map from the historical digitization company Motco Ltd., which Mr. James then set up in MapInfo so that

I would be able to zoom in on the different segments of the map. I spent some weeks visually inputting 287 known trade locations—rather than trade members, since many instrument makers moved one or more times over the course of their independent working lives—according to the streets and other features depicted on the map rather than according to mathematical coordinates. These trade locations were drawn from Gloria Clifton's *Directory of the British Scientific Instrument Makers 1550–1851* and from my own research to that date (Clifton 1995). There were sometimes multiple locations in early modern London which bore the same name, and Rocque's map also did not record all locations or all landmarks such as inns and taverns. Therefore, I frequently had to consult sources such as Henry Harben's *A dictionary of London* and Ralph Hyde's *The A to Z of Georgian London* in order to finalize the points on my map (Harben 1918; Hyde 1981). I did not employ the locations of trade members for which there were no dates provided, since they could not have been incorporated into analyses that reflect the passage of time. In GIS programs, the locations which you pinpoint on the map are tied to databases, so I was able to attach relevant biographical data to each shop site. I originally had grandiose plans of storing all of the information which I collected about trade members in this database but concluded that it would be most time-efficient, in the course of a doctorate, to only input the data most relevant to geographical and time-based analyses of the instrument trade. These included name, known years at work, livery company affiliations (Clockmakers, Spectaclemakers, Grocers, etc.), and job specialties (globe maker, mathematical instrument maker, "multiclass" instrument maker, etc.). The results were more than satisfactory.

7.4 Reaping the Benefits of Vernacular GIS

These efforts initially produced a map of the known locations, rooted in time, of all of the instrument makers and sellers known to have been at work in London during the first half of the eighteenth century (see Fig. 7.2). This revealed that the instrument trade in the capital was more expansive and comprehensive than had previously been understood—extending from St. James's in the west to Shadwell in the east and from Clerkenwell in the north to Southwark and Bermondsey in the south. The highest concentration of locations extended in a rough line from the western end of the Strand along Fleet Street to the area of Ludgate Street and St. Paul's Churchyard and then in clusters appearing further east near the Royal Exchange, the Tower, and the wharves of Wapping beyond the walls of the central metropolis. As could be seen in my ensuing cartographic examinations of specific neighborhoods and of trade specialties, the locations in Wapping represented trade members who mainly targeted customers involved in shipping and trade—for example, compass makers and instrument makers and sellers who were also ship chandlers—while the rest of the concentrations of trade members aligned with the major thoroughfares and shopping areas of the metropolis.

Fig. 7.2 My initial digital map of the members of the London scientific instrument trade who were at work within the period 1700–1750, created with MapInfo (Source of underlying map: Motco Enterprises Ltd., with permission)

While it was very useful to have such a visual representation of the geographical extent and clustering of the trade as a whole, the renowned geographer E.G.R. Taylor similarly mapped the locations of a number of trade members by hand in 1954 (Taylor 1954). What significantly set my efforts apart, and made the time spent setting up the database in MapInfo and creating hundreds of maps worthwhile, was my being able to attach the variable of time and biographical information to each of the geographical locations. With this information incorporated, the initial map made it much easier to track the interrelations and potential competition between trade members in different neighborhoods and streets, since I could swiftly check the details attached to each point on the map. It could also be used to spawn hundreds of further maps to aid in examining different interconnected geographical, socio-economic and time-based aspects of the trade. By "querying" the interactive map—for example, asking it, for example, for only the locations of trade members who belonged to the Grocers' Company and made or sold globes from 1740 to 1745—I generated more than 250 different maps that depicted these trade members and their known locations with respect to different variables and with respect to the passing decades.

By producing maps of the instrument makers at work in London during each of the decades of the first half of the eighteenth century, I was able to chart the increasing expansion of the trade in all directions as time passed, alongside the expansion of the metropolis but especially toward the fashionable west.[18]

[18]See additional discussion of the role of GIS technology in allowing a researcher to create his or her own maps in the chapter in this volume authored by von Lünen and also the chapter by Mares and Moschek.

During the final decade, there was increased clustering in traditional areas as well, including the stretch from the Strand to St. Paul's and the neighborhoods of the Royal Exchange and the Tower, which was probably a result of the continued growth of the trade as a whole. It was also a result of the increasing popularity of the optical and the "tri-class" instrument specialties (the latter selling optical, mathematical, and philosophical wares), which appealed to the affluent customers who frequented most of those locations. While some of these trends may have been expected, given our understanding of the instrument trade and of the nature of demand in different areas, they had never before been proved or analyzed in depth. Incorporating these maps into publications and presentations had a great didactic impact upon readers and listeners as well, by making it easier for them to visually grasp the extent and dynamics of a trade spread across hundreds of shops and diverse urban neighborhoods. A PowerPoint "movie" of the successive decadal maps was particularly well received at a conference since it exhibited, visually and dynamically, how the trade expanded across and clustered in London as time passed.

The maps of the locations of trade members who pursued specific instrument making specialties or combinations thereof revealed settlement patterns that seem to have been shaped by the locations of strongest demand for different types of instruments and by differing rent levels. For example, all but one of the compass makers and sellers were located in Wapping and Ratcliff to the east of the medieval city walls. The east was largely oriented toward the workers and crafts that served shipping and trade and also offered lower rents than could many other areas of London, which would have especially been important to trade members who pursued the typically lower paying instrument specialties such as compass making. In comparison, the makers of globes and timekeepers were mostly located near the Royal Exchange and to the west, as befits the production and sale of instruments that were generally more luxurious and could be displayed in affluent homes as well as being put to practical and educational use. Trade members who pursued one of the publishing trades in addition to instrument making, including map and print selling, mainly congregated in their traditional neighborhoods from the Strand to Fleet Street. Most of the rule makers and sellers settled in an arc from Ludgate Street over St. Paul's Churchyard to Cheapside, most likely because of their ties to the trades in publications and in other mathematical instruments.

As a whole, the optical instrument makers and sellers were mainly concentrated from Temple Bar to Ludgate Street and St. Paul's Churchyard, near the Exeter and Royal Exchanges, and near the Tower, and they increasingly expanded into the west and later somewhat to the east. General optical instrument makers and sellers were also scattered across the neighborhoods of Holborn, west Wapping, and Redriff. The main concentrations of mathematical, philosophical, and tri-class instrument trades shifted westward during this period as well. The mathematical instrument makers and sellers covered the most ground, likely because of the variety of professions and economic classes that purchased and used their instruments and because mathematical instruments were so often sold alongside other goods from maps to fashionable trinkets. The core concentration of mathematical instrument makers and sellers extended from St. James's and especially from Covent Garden east to the Royal Exchange, and then reappeared in clusters in the Minories and near the Tower, with their northernmost point being Moorfields. Their locations expanded in all directions and slightly to the south bank of the Thames over time before largely consolidating in the west, and they reflected the economic range of the specialty from the higher-end neighborhoods between St. James's and the Royal Exchange to the lower-end areas on the periphery.

The philosophical and tri-class instrument trades skewed more toward affluent and learned customers and were thus even more strongly oriented toward the west. The philosophical instrument makers and sellers were located entirely to the west of the metropolis, especially: at St. James's; near Fleet Street and the nearby hospitals; and in Holborn. The tri-class trade only existed in Fleet Street until the last decade of this period. By then, the specialty had increased in popularity or perhaps profitability, and locations also appeared near Piccadilly, Ludgate Street and St. Paul's Churchyard, the Royal Exchange, and the Tower. It is clear that the places in which instrument makers and sellers decided to settle were in large part determined by the types of wares that they sold and the types of customers they courted. This resulted in members of the instrument trade frequently being located near trades that courted similar customers and employed similar skills and materials, including the making of timekeepers and jewelry and the trade in maps, prints, and publications.

However, many other factors could play a role in geographical decision-making as well, including the location of the neighborhoods in which the trade members had been raised or had served their apprenticeship, and the foci of the immigrant or religious communities to which they belonged. My maps also facilitated the study of these factors by making it easy to check the proximity of trade members to their fellows and to key landmarks which either already appeared on Rocque's map or which I added. These could include a Dissenting church, an educational institution, the hall of a livery company, or an important institutional customer for instruments. Such mapping contributed significantly to the otherwise textual study of the diverse socioeconomic factors which influenced the lives and businesses of most trade members. For example, the maps made it clear how much such ties influenced the shop location and business of the mathematical instrument maker Richard Bates (apprenticed 1714–died 1750). Bates had attended Christ's Hospital School, later supplied rules for its Drawing School, and was apprenticed to William Haddon. His master's son was later one of his own apprentices, and Bates' shop facing the Old Bailey was not far from Haddon's shop and Christ's Hospital. In his will, the instrument maker left most of his estate to his wife but also left one guinea for "his friend John Farmer" to buy a mourning ring.[19] Farmer was a mathematical instrument and rule maker whose shop was not in the neighborhood but who had served his apprenticeship nearby at the same time as Bates. Farmer's nephew and successor, Richard Bates Gearing, was Bates' final apprentice, which no doubt strengthened the two men's relationship even more.

[19] Will of Richard Bates, 20 November 1750, The National Archives, PROB 11/783.

The digital maps also assisted my study of other subsets of the trade, whether the customers and business associates of a specific trade member or trade members from a specific immigrant or religious group such as the French Huguenots. At least 5–7% of known instrument makers and sellers seem to have been of French Huguenot descent or to have married into that community, if not more, and it was easy to produce a map of those trade members alone. The French Huguenots formed the single largest, most cohesive and most influential immigrant group in the instrument trade of this period, being bound together by shared persecution, blood, language, and communal institutions including churches, hospitals, and charities (Scouloudi 1987; Gwynn 1985). Since this community could offer many contacts in the luxury metalworking trades, it seems to have been relatively common for instrument makers who married into it to expand their businesses to include wares like toys and jewelry as well as instruments. The maps assisted me in tracing the geographical and socioeconomic attributes of this dynamic and in seeing what competition these individuals likely faced from within the trade.

The instrument trade's intersections with the trades in toys and other luxuries mainly took place in the central City of London and on the western side of the capital, since those were the main centers of fashion and high-end retail shopping during this period. All but one of the French Huguenot individuals or partnerships who practiced this combination of trades seem to have lined the important retail corridor of Fleet Street, Ludgate Street, and St. Paul's Churchyard—as did other members of the trade who sold similar combinations of wares. MapInfo also made it easy to map and thus to geographically analyze relevant individuals outside of the instrument trade, as in the aforementioned case of the London customers and business associates of the toyman and optician George Willdey. I created a separate basic database of these individuals, which allowed me to chart their locations on Rocque's map, and with respect to Willdey's shop location, while ignoring the other members of the instrument trade. The results emphasized how commercial traffic moved throughout the early modern metropolis, and how the socioeconomic networks of relationships upon which an instrument business was often based were embedded in the landscape of London.

In addition to these more innovative uses of digital mapping as a representational and analytical tool, I was of course able to consult the basic map of all trade members as one would a traditional version—to judge whether a given neighborhood, street, or court might have supplied the types of buildings which an instrument maker would require for production or for attracting passing custom. They required enough sunlight and sufficient space for the conduct of their trade specialty, with the makers of large mathematical and astronomical instruments needing far more space for their work and storage than did most spectaclemakers and opticians. For a business that involved retail sale, it was best to be positioned on or as close as possible to a major thoroughfare, one of the more fashionable squares or a prominent landmark such as St. Paul's Churchyard or the Royal Exchange. Some shop-owning instrument makers operated in courts, alleys, and yards, but the more distant they were from passing trade, the more effort they had to invest in attracting potential customers. Although I did not go to such lengths, one could also enhance such digital maps with related data like the local rent levels, drawn from the rate books. A single neighborhood or even a single street might encompass a range of rents due to different types of housing and to the positioning of buildings, but there were still differences in the overall rent levels to be found in different parts of the metropolis.

Since GIS programs can be used to zoom in on or out from individual streets and neighborhoods, as well as to look at the overall picture, the technology lends itself to the study of the nature of different neighborhoods and their inhabitants. Each area of London had its distinctive personality resulting from factors such as the trades, institutions, and socioeconomic nature of individual located therein. However, my digital maps of the instrument trade reinforced that there was a good deal of traffic between and interrelationships connecting the different parishes and neighborhoods, as well as many similarities. As historians including Derek Morris have emphasized in recent years, the eastern side of London above the Thames was not simply a crumbling warren of houses fit only for roughhousing sailors as has sometimes been represented, and the west was not an idyllic and orderly grid of mansions that only housed the titled and the *nouveaux riches* (Morris 2002; Morris and Cozens 2009). Varying proportions of the poor, servants, laborers, shopkeepers, craftsmen, professionals, intellectuals, the wealthy, and different institutions inhabited both sides of the capital north of the Thames. However, there were real socioeconomic differences between many of the parts of the metropolis, and these were especially dramatic between east and west and produced significant differences in the nature of the instrument trade in both locations.

The west was, broadly speaking, a much more affluent, fashionable, and politically and socially influential area. The east was mainly oriented toward shipping and related crafts and trades, as well as to larger scale manufacturers. These differences are reflected in the instrument trade specialties practiced in each location, with the trade members in the west containing a large proportion of optical instrument makers and spectaclemakers, in addition to a majority consisting of mathematical instrument makers including globe makers and sellers. They also encompassed some makers of timekeepers, a number of toymen and jewelers, and most of the trade members who sold philosophical instruments or all classes of instrument. In comparison, almost all of the trade members on the eastern side of the metropolis were mathematical instrument makers and sellers, including a number of rule and especially compass makers who did not appear further west, and they were sometimes ship chandlers as well. Sources including Wills and Sun Insurance policies and records of royal and institutional patronage show that this resulted overall in greater wealth being accumulated, and more public acclaim being garnered, by the trade members in the west than in the east.

As can be seen through all of these examples from my doctoral research, GIS mapping, as Knowles and Hillier have said, "offers an unprecedented range of tools to visualize historical information in its geographic context, examine it at different scales, interrogate its spatial patterns, and integrate material from many sources on the basis of shared location" (Knowles and Hillier 2008, xii). It provided me with very useful maps of the entire instrument trade and of a myriad of subsets of said trade, such as the trade members who belonged to a specific livery company during a specific decade. It would have taken much longer to produce hundreds of variable-based maps using traditional methods than it did using GIS, and the traditional versions would not have been interactively tied to other data. With these maps, it was much easier to trace the geographical and socioeconomic dynamics for hundreds of shops and workshops and at different scales. They aided me in tracing the ties and competition which existed between different trade members, and between trade members and other individuals, institutions, and geographical features. Since GIS provides an interactive rather than a static system of mapping, I could also generate new maps in the future based upon my original sample or add additional individuals and types of information. Furthermore, I can easily turn these maps into evocative digital or print images for sharing my research with others. Tools such as these can do much to help historians to avoid the tendency to examine the past "as if it were packed solidly on the head of a pin, in a fantasy world with virtually no spatial dimensions"—as Edward Soja famously first commented about Marshall and Pigou's view of the economy (Soja 1989, 32).

7.5 Addressing Concerns About the Use of GIS in Historical Research

My experiences with digital mapping show that employing vernacular rather than precise GIS maps can answer many of the key concerns about applying such technology to early modern historical research. One of the most common objections is that GIS cannot take into account the variable and incomplete nature of sources from the early modern period.[20] There are often gaps in the individual and institutional records which were produced or have survived from before the nineteenth century, as well as great variability in their accuracy and degree of detail. Some geographers and historians have investigated different methods of representing uncertain or incomplete information on a map in order to overcome these obstacles but so far with little success. The majority perception of maps remains that they are dependent upon data sets being accurate and complete, which is seldom the reality for pre-modern material. GIS technology can perpetuate this misconception since, as Anne Kelly Knowles and Amy Hillier have said, it "tends to reinforce the naïve acceptance of maps as authoritative statements because the software so swiftly produces maps behind whose veneer of professionalism may lie all manner of unseemliness" (Knowles 2008, 19).

Some proponents of the use of GIS in the humanities have suggested overcoming this problem by using multimedia approaches to make the technology "more fluid and ambiguous," but so far, the more successful response seems to be to avoid unnecessary geographical precision whenever possible (Corrigan 2010, 76). In terms of working with gaps in the data, research into early modern populations is almost

[20]Gregory and Ell (2007, 1), Corrigan (2010, 76), Knowles (2008, 3) and Pickles (1999).

always a matter of combining individual stories with the best-possible statistical analyses of the surviving evidence about the whole, since records from this period are almost never complete nor homogeneous. Such records can be incorporated into digital maps as easily as they can into these case studies and statistical analyses, as long as they have some geographical component and ideally (but not necessarily) the added variable of time. If a certain attribute is not known for some individuals—for example, in my case, to which livery company an instrument maker belonged—then it can simply be set to "unknown" rather than invalidating the incorporation of that individual. If the information on some individuals is less precisely known than for others—for example, their location only being known to the level of a city parish rather than to the street level—then it can simply be provided with a different symbol or color on the digital maps and included or omitted at will. As you can see in Fig. 7.2, I utilized green squares rather than the usual red circles to delineate which of the trade members on my maps were only located according to parish. If the information displayed is generally a bit questionable, such as the exact dates of employment for some individuals, then this is equivalent to the uncertainty reflected in the textual and statistical analyses of the same sample; one could choose to add a field which notes the existence of this ambiguity. For this period, digital maps and case-based and statistical analyses of a population can all reveal key dynamics despite any imprecision at the individual level.

Similar concerns have been raised about the precision of GIS base maps, since the maps and other geographical records of European cities were not very accurate until at least the later 1700s if not the 1800s. To address this issue, Craig Spence combined information from multiple maps of London for his research into the late seventeenth-century capital, noting, "[t]he quality and detail of seventeenth and eighteenth-century maps is both varied and unpredictable, furthermore none meet modern surveying criteria for accuracy" (Spence 2000b, 34). These included Morgan's map of the City and Westminster from 1682, the parish maps printed in Strype's *Survey of London* in 1755, Rocque's map, other local maps to define the boundaries of the original assessment districts, and the Ordnance Survey Maps of the 1860s. In working with GIS, there are additional issues of the time and cost involved in attempting to scan a map at a high enough resolution to allow zooming in and out of the landscape and sometimes in joining up the individual sheets of a multipart map, and with precision GIS, there is also the complication of connecting the map to modern GIS coordinates with georeferencing techniques.

As previously discussed, this level of precision is unnecessary for most early modern research because of the ways in which contemporaries typically perceived and depicted their surroundings and because of the small geographical spaces inhabited by urban populations before the modern era. If, for example, one scientific instrument shop was within sight and easy walking distance of another on a London street, why would the precise distance between them really matter? The important point to be considered when you want to employ digital mapping is to find an early modern or contemporary depiction of the main features and thoroughfares of the location that you are examining, which reflects their basic spatial relationship to each other, and upon which you can roughly locate the entities in your sample.

Choosing the map and geographical features which can best accomplish this without incorporating precise mathematical coordinates is similar to the process of "satisficing" which Hill describes with the respect to precision GIS. This involves deciding which shapes, from points to multisided geometric figures, are sufficient for approximating each of the features being mapped without losing time and money to unnecessary attempts at optimization. The example which the author provides is of a point often being sufficient to represent an individual or a "point-like" entity such as a monument but often not being sufficient to represent the position of a river with respect to neighboring features (Hill 2000, 288–289). If a suitable early modern depiction of your target area does not exist or cannot easily be scanned at a high enough resolution, then it may be possible to simply produce a modern drawing of the basic streets and landmarks. This provides a platform for representing the general spatial arrangement of the locale, and perhaps the inhabitants' understanding of their surroundings, upon which you can build an incredibly useful visualization and analytical tool for your overall research—even if it is not as aesthetically pleasing as an early modern map.

The use of vernacular mapping can overcome many other common objections to the application of GIS to the study of the humanities as well, including to the cost and labor and time demands of such an enterprise.[21] Relevant programs are often made freely or affordably available through different institutions or online, and as previously discussed, vernacular mapping seldom requires as much effort being spent on producing the base map as does precision mapping. Creating the databases for the maps is considered another costly step in using GIS, with some authors estimating that it accounts for up to 85 % of the total cost (Longley et al. 2005, 201). However, it should cost no more than it would to construct other databases for collaborative research, and in the case of studies of smaller populations such as my hundreds of instrument makers, it can be accomplished by a lone researcher. It is often stated that GIS is a tool best used by a research team rather than by an individual due to not only cost but also to the time investment needed to create the associated databases and due to the diverse skills involved in exploiting a geohistorical technology (Gregory and Ell 2007, 11). However, this is clearly not the case with vernacular mapping that does not incorporate great geographical precision, since it can be set up by one scholar or student (in my case with initial assistance from a Map Librarian) as long as they restrict the amount of information which they intend to input.

Finally, some sources warn that researchers and students can misuse technology like GIS, such as by poorly interpreting the maps produced with it or by viewing them as an end in themselves rather than as a tool for aiding in rigorous academic scholarship.[22] The same concern was raised before I began the digital mapping of the London instrument trade for my doctorate, and I have since heard it repeated at a number of GIS-related meetings and lectures. I would argue that this is not a problem that is specific to GIS but simply a potential complication of

[21] Gregory and Ell (2007, 17, 41, 89), Martí-Henneberg (2011, 11–12), Knowles (2000) and Siebert (2000).

[22] Gregory and Ell (2007, 1, 12) and Knowles and Hillier (2008, 19).

historical research. If a scholar applies the proper amount of consideration and contextualization to their sources and to other forms of analyses, then there is no reason for them to not do the same with their digital maps and information. In my own research, the maps were always consulted in conjunction with a wide range of primary sources, statistical analyses, and case studies.

For example, they were very useful in conjunction with documentary sources in analyzing the scientific instrument trade cluster which existed during the early 1700s in Wapping on the eastern side of London, near the bridge over the Hermitage Dock and near the Hermitage Stairs close by. The maps made it clear what a pronounced cluster there was at that location and how close it was to the Hermitage Dock, which was the southwestern entrance to the London wharves, and to works including the Hermitage Pothouse and wharves for coal and timber. They also aided me in examining the interrelationships of the trade members located there and their trade specialties, which were mostly river- and sea-oriented. However, it was the diverse primary sources that began to fill in the large gaps which existed in the history of this area and exposed the complicated networks of commercial and livery company associations and of blood and marriage which bound most of these men together. Statistical analyses then allowed me to compare the nature and affluence of this cluster to those of trade clusters in other parts of the metropolis. GIS and other digital technologies are useful tools, while it is the resources, analyses, and publications which emerge from them that are the actual outputs of the research.

A valid concern for the usage of GIS is planning, which of course applies to most elements of research and to the production of any digital technology such as standalone databases. The creation of a map depicting a significant population will take some time, even if it does not require the achievement of great geographical precision or the inputting of very many database fields (i.e., name, dates, livery company, trade specialty). It is therefore vital on a set-length project for a researcher or research team to decide as early as possible which types of the following would be most useful and time-efficient for the research being conducted: mapping program; base map; sample size and makeup; and database fields. It is often difficult to predict at the outset of a project what will prove most useful, and one can always add additional database fields and sample members at a later date. However, an early and organized start can keep a GIS mapping initiative a valuable analytical and visualization tool for historical research rather than a hindrance.

Digital mapping is a technology which is only becoming more accessible and less expensive as the years pass. It holds untold promise for contributing to the analysis of historical populations and events which are firmly rooted in their geographical and socioeconomic contexts. It is also continues to facilitate the sharing of observed dynamics and research conclusions with other researchers and with the general public. One interesting result of this has been the publication of social atlases like those of Spence, Woods and Shelton, and Kennedy et al.[23] However, the future will likely see a greater move toward the electronic publication and hosting of such

[23] Spence (2000b), Woods and Shelton (1997) and Kennedy et al. (1999).

maps and toward their incorporation in electronic databases and libraries, in order to allow users to interact with them in full and at lower cost (Gregory and Ell 2007, 11, 145–160). While a research team would typically be required to produce such resources when precision GIS is involved, the usage of vernacular maps makes it likely that even individual researchers and students will be able to contribute maps and databases of early modern populations to such an interconnected digital future.

References

Arias S, Warf B (2009) Introduction: the reinsertion of space into the social sciences and humanities. In: Arias S, Warf B (eds) The spatial turn: interdisciplinary perspectives. Routledge, London

Baker A (2006) The London instrument trade, from Culpeper to Cole. In: Grob B, Hooijmaijers H (eds) Who needs scientific instruments? Boerhaave Museum, Leiden, pp 99–105

Baker A (2009a) The business of life: the socioeconomics of the 'scientific' instrument trade in early modern London. In: Eliassen FE, Szende K (eds) Generations in towns: succession and success in pre-industrial urban societies. Cambridge Scholars Publishing, Newcastle upon Tyne, pp 169–191

Baker A (2009b) Reading between the lines: the instrument trade in the newspapers of 18th century London. Sci Instrum Soc Bull 102:12–17

Barnett G (ed) (1936) Two tracts by Gregory king. Johns Hopkins, Baltimore

Beier A, Finlay R (eds) (1986) London 1500–1700: the making of the metropolis. Longman, London

Berry H (2002) Polite consumption: shopping in eighteenth-century England. Trans Royal Hist Soc Sixth Ser 12:375–394

Bunge W, Sack R (1973) Spatial prediction. Ann Assoc Am Geograp 63(4):566–568

Butlin R (1993) Historical geography: through the gates of space and time. Hodder Arnold, London

Byrd M (1978) London transformed: images of the city in the eighteenth century. Yale University Press, New Haven

Clifton G (1995) Directory of British scientific instrument makers 1550–1851. Philip Wilson, London

Corrigan J (2010) Qualitative GIS and emergent semantics. In: Bodenhamer DJ, Corrigan J, Harris T (eds) The spatial humanities: GIS and the future of humanities scholarship. Indiana University Press, Bloomington, pp 76–88

Curran M (2010a) 'Mettons toujours Londres': enlightened christianity and the public in pre-revolutionary francophone Europe. Fr Hist 24(1):40–59

Curran M (2010b) The Société typographique de Neuchâtel and networks of trade and translation in eighteenth-century francophone Europe. In: Thomson A, Burrows S, Dziembowski E (eds) Cultural transfers: France and Britain in the long eighteenth century. Voltaire Foundation, Oxford, pp 257–267

Davis D (1966) A History of shopping. Routledge, London

Field J (1988) What is scientific about a scientific instrument? Nuncius 3(2):3–26

George M (1925) London life in the eighteenth century. Harmondsworth, London

Gregory I, Ell P (2007) Historical GIS: technologies, methodologies and scholarship. Cambridge University Press, Cambridge

Guillery P (2004) The small house in eighteenth-century London: a social and architectural history. Yale University Press, London

Gwynn R (1985) Huguenot heritage: the history and contribution of the Huguenots in Britain. Routledge, London

Harben H (1918) A dictionary of London. Herbert Jenkins, London

Hill LL (2000) Core elements of digital gazetteers: placenames, categories, and footprints. In: Borbinha J, Baker T (eds) Research and advanced technology for digital libraries. Lecture notes in computer science, vol 1923. Springer, Berlin/Heidelberg, pp 280–290
Hill L (2006) Georeferencing: the geographic associations of information. MIT, Cambridge
Hyde R (ed) (1981) The A to Z of georgian London. Harry Margary, Lympne Castle
Jeffreys J (1954) Retail trading in Britain 1850–1950. Cambridge University Press, Cambridge
Kennedy L, Ell P, Crawford E, Clarkson L (1999) Mapping the great Irish famine, an atlas of the famine years. Four Courts Press, Belfast
Knowles AK (2000) Introduction. Soc Sci Hist 24(3):451–470
Knowles AK (2008) GIS and History. In: Knowles and Hillier (2008), pp 1–25
Knowles A, Hillier A (eds) (2008) Placing history: how maps, spatial data, and GIS are changing historical scholarship. ESRI, Redlands
Kremer R (2007) 'Making one of Saturn's moons': instrument making along new England's northern frontier. Presented at the 26th symposium of the scientific instrument commission, Cambridge, MA
Langton J (1972) Systems approach to change in human geography. Prog Geogr 4:123–178
Longley P, Goodchild M, Maguire D, Rhind D (2005) Geographical information systems and science, 2nd edn. Wiley, Chichester
Martí-Henneberg J (2011) Geographical information systems and the study of history. J Interdiscip Hist 42(1):1–13
Massey D (1999) Space-time, 'science' and the relationship between physical geography and human geography. Trans Inst Br Geogr New Ser 24:261–276
Massey D (2005) For space. Sage, London
Merritt J (ed) (2002) Imagining early modern London: perceptions and portrayals of the city from Stow to Strype, 1598–1720. Cambridge University Press, Cambridge
Morris D (2002) Mile End Old Town, 1740–1780: a social history of an early modern London suburb. East London History Society, London
Morris D, Cozens K (2009) Wapping 1600–1800: a social history of an early modern London maritime suburb. East London History Society, London
Morrison-Low A (2007) Making scientific instruments in the industrial revolution. Ashgate, Aldershot/Burlington
Ogborn M (1998) Spaces of modernity: London's geographies, 1680–1780. Guilford, New York
Pepys S (2006) The diary of Samuel Pepys: 1668. Hard, Teddington
Pickles J (1999) Arguments, debates, and dialogues: the GIS-social theory debate and the concern for alternatives. In: Longley P, Goodchild M, Maguire D, Rhind D (eds) Geographical information systems: principals, techniques, management and applications, 2nd edn. Wiley, Chichester, pp 49–60
Power M (1986) The social topography of restoration London. In: Beier A, Finlay R (eds) London 1500–1700: the making of the metropolis. Longman, London, pp 199–223
Raven J (2004) St. Paul's precinct and the book trade to c. 1800. In: Keene D, Burns A, Saint A (eds) St. Paul's: the cathedral church of London 604–2004. Yale University Press, London, pp 430–438
Rudé G (1971) Hanoverian London 1714–1808. Secker and Warburg, London
Sack R (1972) Geography, geometry, and explanation. Ann Assoc Am Geogr 62(1):61–78
Sack R (1974) Chorology and spatial analysis. Ann Assoc Am Geogr 64(3):439–452
Scouloudi I (ed) (1987) Huguenots in Britain and their French background, 1550–1800. Barnes and Noble, London
Siebert L (2000) Using GIS to document, visualize, and interpret Tokyo's spatial history. Soc Sci Hist 24(3):537–574
Soja E (1989) Postmodern geographies: the reassertion of space in critical social theory. Verso, London
Spence C (2000a) Computers, maps and metropolitan London in the 1690s. In: Woollard M (ed) New windows on London's past: information technology and the transformation of metropolitan history. Association for History and Computing, Glasgow, pp 25–45

Spence C (2000b) London in the 1690s: a social atlas. Centre for Metropolitan History, Institute of Historical Research, London
Stobart J, Hann A, Morgan V (2007) Spaces of consumption: leisure and shopping in the English town, c. 1680–1830. Routledge, Abingdon
Taylor E (1954) The mathematical practitioners of Tudor and Stuart England, 1485–1714. Cambridge University Press, Cambridge
Walsh C (1995) The design of London Goldsmiths' shops in the early eighteenth century. In: Mitchell D (ed) Goldsmiths, silversmiths and bankers: innovation and the transfer of skill. Alan Sutton, Stroud, pp 96–111
Warner DJ (1990) What is a scientific instrument, when did it become one, and why? Br J Hist Sci 23(1):83–93
White R (2008) Foreword. In: Knowles A, Hillier A (eds) Placing history: how maps, spatial data, and GIS are changing historical scholarship. ESRI, Redlands, pp ix–xi
Woods R, Shelton N (1997) An atlas of Victorian mortality. Liverpool University Press, Liverpool
Wrigley E (1987) People, cities and wealth. Blackwell, Oxford

Chapter 8
Mapping Miracles: Early Medieval Hagiography and the Potential of GIS

Faye Taylor

8.1 GIS and Medieval Studies

Medieval studies have benefited from recent developments in digital humanities, and the application of GIS to medieval studies is not new. In this area, archaeologists and some historical geographers have paved the way, although to an extent these are misleading definitions—disciplinary boundaries are readily crossed and blurred, and medieval studies are intrinsically multidisciplinary. The technical skills needed to study the period and its evidence make it particularly conducive to collaborative projects that bring together a range of expertise. To date, some of the most exciting work conducted using GIS for the medieval period has arisen from such associations, which bring together historians, palaeography specialists, linguists, geographers, and computer scientists.[1]

Utilisation of GIS in medieval research projects has tended to fall into three broad categories. The first—investigating rural and urban landscapes—is influenced heavily by historical geography and archaeology. The history of medieval rural and urban landscapes has been particularly revealing in terms of land usage and town planning. Two multidisciplinary projects, *Mapping the Medieval Urban Landscape*, exploring aspects of Edward I's new towns in England and Wales, and *Mapping Medieval Chester*, based at Queen's University Belfast, have broken new ground in this area. Both synthesise expertise from archaeology, geography, literary studies, and history to improve our understanding of medieval urban planning and conceptions of space, and the role of space in forming identities.[2] Large-scale

[1]I thank Julia Barrow and Keith Lilley for their comments on an earlier version of this chapter.

[2]http://www.qub.ac.uk/urban_mapping/index.htm and http://www.medievalchester.ac.uk/about/, both last accessed March 2012. The resulting publications particularly by Keith Lilley give

F. Taylor (✉)
Cumberland Lodge, Windsor, UK
e-mail: fayetaylor@cumberlandlodge.ac.uk

A. von Lünen and C. Travis (eds.), *History and GIS: Epistemologies, Considerations and Reflections*, DOI 10.1007/978-94-007-5009-8_8,

collaborative projects, especially those with a regional outlook, are increasingly likely to embrace a GIS element, including the forthcoming *Sherwood: The People of the Forest, 1218–1835* which will plot the changing ecological, cultural, and religious landscape of Sherwood Forest. A better understanding of the landscape can be applied in many ways, including its implication for military logistics, such as those in the Battle of Manzikert (1071) analysed by the Medieval Warfare on the Grid (MWGrid) project.[3]

The second category of GIS use is in cartographic and manuscript analysis, for which the spatial dimension can be twofold. Absolute spatial data is used first to refer to a point on a physical object, allowing the minutiae of original manuscripts to be analysed closely and instructively. The *Lancelot Project* uses GIS in this way (and in others), for analysing medieval manuscripts relating to the search for the Holy Grail. The project team boasts specialists in Old French, art history, and manuscript studies and, in its own words "treats the manuscript page as a conceptual map". GIS is used to "associate information objects with the appropriate passages, illuminations and pictures in the manuscript, to assist the reader or analyst with exploring the manuscript further", opening up a world of interrelated images and text passages to researchers.[4] If the manuscript also happens to be a map, GIS can also be used compare and to analyse the relational representation of space in medieval maps with their modern counterparts. The *Gough Map Project* leads the way in this respect. It was able to conclude that the map was designed to be used in a practical sense and demonstrated a greater accuracy of representation for the southeast of Britain, which suggest an origin for the manuscript itself.[5] In this way, research feeds back in to our understanding both of the cartographic process and medieval concepts of space, in addition to offering solutions to research questions about the manuscript.

The strengths of the first two categories of GIS utilisation lie in the duality of the mapping process and its cartographic outputs. The third area where GIS has been applied to medieval studies projects is more nebulous. It involves the creation of large-scale maps that display a range of different layers of data. This allows the visualisation of various social and political phenomena, together with changes over time and space. The resulting e-resources are akin to traditional historical atlases, but they offer much greater flexibility, accessibility, and potential for public engagement than traditional historical cartography. For example, the *Digital Atlas of Roman and Medieval Civilization* (DARMC) includes information of interest to a wide audience, including

> Roman road networks, bridges, aqueducts, the cities and settlements of the empire, Roman military installations, the shrines, mines, and villas [...] the main kingdoms, empires, and

---

examples of the implications of the research, for example, Lilley (2009). On the field of HGIS from a historical geographer's perspective, see Lilley (2011a) and Lilley (2011b).

[3]http://www.cs.bham.ac.uk/research/projects/mwgrid/, last accessed March 2012.

[4]http://www.lancelot-project.pitt.edu/lancelot-project.html, last accessed March 2012.

[5]http://www.goughmap.org, last accessed March 2012, on which see, for example, Lilley and Lloyd (2009) and Lloyd and Lilley (2009).

> principalities of medieval Europe, the offices of the great trading federation of the Hanseatic League, the spread of Cluniac monasticism, the medieval universities, the transmission of the Black Death, [...] the episcopal sees of the late Roman empire, Byzantium, and medieval Europe; Europe's ecclesiastical provinces and the dioceses of medieval France.[6]

The DARMC makes no claim to definitiveness, and evidently, its ambitious scope relies on a huge input from multiple contributors—in this case, via a pseudo-crowdsourcing model—utilising undergraduates and postgraduates at Harvard. Similarly, *Regnum Francorum Online* (RFO) aims to provide interactive maps containing a wealth of plotted primary source material.[7] Such resources are a useful and accessible reference for any number of individual scholars and the process of visualising patterns over large scales and time periods will surely raise further research questions. The beauty of GIS-based online maps is that each point can be interrogated for its background data source, allowing a transparency of evidential support that facilitates a constructive review process. This, in turn, facilitates constant revision as interpretation changes. Yet this form of mapping is the most susceptible to criticisms founded in the spatial turn[8] and uses GIS as a tool for creating reference works rather than as research tool in and of itself.

All of these three areas of GIS usage promise great potential in more traditional fields of social, political, and religious history of the early medieval period, including for individual historians without the backing of large funding grants or specialist collaborators. Although lacking the extensive economic survey and administrative records of the later medieval period, land in one form or another is central to the majority of surviving evidence from the early and middle medieval era, since property underpinned all power relations.[9] However, the majority of evidence—often dry and formulaic—has hitherto not been subjected to any broad systematic spatial analysis. Historians familiar with the early medieval period are keenly aware of the bias in source survival towards numerous legal documents. These usually comprise accounts of transfers, exchanges, or sales of land, often involving religious institutions, and are sporadic and inconsistent in their creation and survival across time and place. In line with broader developments in the digital humanities, the established field of diplomatic has evolved a digital arm in recent years, reflecting a burgeoning interest in computer-based analyses which allow a great quantity of disparate texts to be investigated in innovative ways. New technologies allow these rich resources to be approached from a variety of angles, with particular headway being made in linguistic/lexicographical analysis. New and exciting collaborative projects promise to build on this momentum, including ChartEx (charter excavator), which will use modern language processing techniques to develop software that could potentially be used with any medieval charter.[10]

---

[6]http://darmc.harvard.edu/icb/icb.do, last accessed March 2012.

[7]http://www.francia.ahlfeldt.se/index.php, last accessed March 2012.

[8]On which, see Layers (2010, 1–3).

[9]Good introductions to this are Wickham (1994) and Davies and Fouracre (1995).

[10]http://www.york.ac.uk/history/history-in-action/digging-into-data, last accessed March 2012.

Since charter-type sources generally concern property, they contain much spatial information. The LangScape project draws on this, providing a searchable database of Anglo-Saxon estate boundaries.[11] A parallel revolution to digital charter analysis has not yet been fully realised in the spatial dimension, however. This is despite the potential that GIS technologies offer to combining the translation of existing databases with traditional mapping, which has long played a central role in medieval history as the wider discipline. A key function of the type of software ChartEx hopes to develop will enable the extraction of targeted word forms, including proper names. Such data mining capabilities will provide rapid access to large banks of spatial data, within the parameters of any number of collections of texts. Nevertheless, a place name is not the same as a coordinate, and there are obstacles to be overcome in locating places for which relational information is not always known. This also applies to the spatial information found in many other genres of narrative textual evidence that medievalists rely on, sources less suited to broad computer-based analyses, although they are often used together with legal documentation. Narrative sources require different types of linguistic and literary analysis and are often analysed independently and in isolation. They nevertheless demonstrate similar concerns with land, geography, and space.

There are plenty of opportunities for individual historians to benefit from the use of GIS in their research, therefore, particularly in its cartographic manifestation.[12] The capabilities and applications of GIS are widespread but depend on the questions posed. The ability to visualise data, particularly as maps, is perhaps the most immediately evident advantage to the individual historian with no prior GIS experience. Cartography remains an art and publication-standard maps need not be the goal: GIS makes cartographic and geovisual analysis more accessible and allows maps to be used by the historian as research tools in their own right.[13] To the uninitiated, GIS technologies and techniques can appear impenetrable, perhaps even irrelevant. However, written from the perspective of a non-specialist in GIS, this chapter explores some of the reasons why social and political historians of the early medieval period may wish to gain familiarity in them and some methodological concerns that the process raises. It does so by outlining two example cases that explore the medieval miracle collections of Saints Colombano and Foy. In these examples, GIS has or could suggest alternative interpretations to existing research questions.[14]

---

[11] http://www.langscape.org.uk/index.html, last accessed March 2012.

[12] Individual historians are beginning to use GIS for making their own maps, for example, Armstrong (2008, 128); see also the chapter by Alexander von Lünen and the interview with Gunnar Olsson in this volume.

[13] As pointed out in Gregory and Ell (2007, 10).

[14] The broader context and further references for the arguments made herein are to be found in Taylor (2012).

8.2 San Colombano di Bobbio and Holy Journeys

The demise of the powerful Carolingian Empire was followed by a series of transitory monarchies in France and Italy, with much-debated consequences for society.[15] At the same time, saints' cults were promoted with urgency as religious institutions—in these cases, monasteries—sought to broaden their influence and wealth. Literature concerning saints (collectively referred to as hagiography) and especially collections of posthumous miracle stories (known as *miracula*) offer a unique narrative perspective on these social and political issues in supplement to charters of land transfers.[16] Technologies that allow the investigation of spatial concerns are particularly interesting because the conflicting and ever-changing nature of different forms of power and power centres, with different "symbolic geographies", has been broadly recognised (Theuws et al. 2001). Recent historiographical developments have focussed on ideas of "spatial logic", in particular the evocation and representation of "territories", and, for the religious landscape, on the construction and materialisation of space (representational and otherwise) around a religious institution.[17] Despite the restrictions of the genre, miracle collections vary widely in their length and in the way that their compilers intended them to be read or heard. Accordingly, they represent spatial landscapes in different ways, and analysis of the spatial landscapes evoked by the miracle texts has much to tell us more about how each monastery sought to engage with the sociopolitical landscape. As a result, GIS technologies offer quite different opportunities for each study.

Unlike the Sainte-Foy miracle stories which contain geographic data in an incidental manner, the miracles of San Colombano di Bobbio place a journey at their centre and thus evoke space more explicitly. Colombano was an Irish missionary monk who travelled across Europe and founded several monasteries, the last of which was Bobbio, in the Apennines in north-west Italy. Bobbio's foundation was partially due to a political strategy by the Lombard king, Agilulf, who granted Colombano the land on which to build it, equidistant from both the dioceses of Tortona and Piacenza. The bishops of these two sees quarrelled for centuries over jurisdiction in the area and over Bobbio. An anonymous member of the Bobbio community wrote a collection of *miracula* about Colombano in the mid-tenth century.[18] There miracle stories, unlike the Sainte-Foy collection, were intended to be read as one work, comprising only twenty-eight chapters. It seems probable

[15] In particular, the debate referring to "feudal revolution", which began with the seminal study of Duby (1953) and exchange in the pages of *Past and Present*, sets many of the parameters of the debate: Bisson (1994), Barthélemy and White (1996) and Reuter and Wickham (1997). Literature on different components of the complex paradigm presented by Duby is now vast.

[16] The rhetorical nature of medieval hagiographical texts is now fully acknowledged, with terms such as "monastic weapons" and "propaganda" applied to them: Rosenwein et al. (1991, 765), Ward (1987, 2) and Rollason (1985, 90).

[17] A concept originally advocated in Lauwers (2005). See also Devroey and Lauwers (2007).

[18] Bresslau (1934), henceforth *MSC*.

that the miracle stories were intended to be heard or read by a king, perhaps Otto I in the 960s, to whom the monks directed a plea for the restoration and protection of their land. Monastic arguments drew on a comparative dispute three decades earlier, when the central events recorded in the collection occurred, by making reference to an unnamed bishop. The middle chapters of the text concern a procession of the saint's relics in 929 from Bobbio to King Hugh and his royal court at Pavia, then back. This holy journey formed part of an attempt to confirm a royal and public sanction for the recovery of usurped lands and to stake a claim to the area Bobbio claimed as its territory. The journey took almost three weeks in all, including a few nights' stop in Pavia, and the miracle stories provide some information about where and when the procession stopped along the way. Charter evidence confirms that these way stations, almost exclusively, were plots of land Bobbio already had an interest in. Identification of some of the places from their Latin names is not without debate, and as a result, two possible routes have been proposed by Tosi (1981) and Nuvolone (2000).

Initially, these routes were mapped in order to analyse the extent of land claimed in the region, both by local diocesans and the monastery of Bobbio. However, during the process a more sophisticated use of GIS came to light. The cost analysis function may help answer whether or not the description of the procession was literal or stylised, that is to say, whether it was feasible in the time and schema stated. It could also suggest better which route was taken. We know various factors that could be used for the equations: the procession took place in July 929 (this could be integrated with climate projections); it was attended by monks, including older members of the community, as well as the local lay community; they were carrying a heavy pine chest containing the relics whilst ringing bells and singing. The route begins high in the Apennines and drops down and across the river Po. The area, devoid of major tectonic and volcanic activity, is unlikely to have changed significantly in its topography since the medieval period. A cost analysis may allow for the likely route to be established, given the timings and abilities of the procession indicated by the miracle stories. An alternative or complementary method would be to use a viewshed analysis. This gives users a view of what would be visible if standing on a given spot on the map and facilitates speculation on the layout of roads through the valleys on which the historical sources are silent.[19]

Establishing the route may, in turn, help clarify the conceived territorial bounds of the monastery in the first half of the tenth century. It is likely that the journey was an important statement about ownership, territory and power since the miracles also tell us that trees were carved with crosses along the route, fulfilling a boundary-marking function.[20] More generally, it was a statement of territorial assertion directed via a tirade in the miracle stories (ref. MSC, c. 23) towards a local bishop—either the bishop of Tortona and Piacenza—who was seeking to extend and assert his episcopal influence in the region at that time. If, for example, the procession

[19]"Viewshed analysis" is a standard function to be found in most GIS packages.

[20]A practice also known elsewhere on the continent: Dreslerová and Mikuláš (2010).

went as far east as Sarturano (Agazzano) as Tosi believes, the message was directed to the Bishop of Piacenza (which lies to the north east)—it would mean that Bobbio considered its land to extend far into the same area the bishop claimed as his. The same applies to the diocese of Tortona, to the west. Such local power relations, and their timeliness, are of particular importance in this period. The king was weak, local power bases were frequently redefining themselves, and borders were being established. Establishing which diocesan was considered a threat at that time would tell us about the bishops themselves and may also help answer the mystery surrounding the unidentified diocesan subjected to the tirade in the miracle stories.

8.3 Political and Cultic Spheres of Influence at Sainte-Foy de Conques

The Sainte-Foy miracle collection is of a different nature to that of Colombano. Martyred as a child during the Roman era, Sainte Foy was renowned from the ninth century when the monks of Conques established a cult centre to her in the medieval county of the Rouergue (dép. Aveyron, southern France). From the early eleventh century work began on redacting stories of posthumous miracles attributed to Sainte Foy. Today, there remain 120 discrete miracle stories dating to the eleventh century, written by three or more authors presented in four books and further additional accounts.[21] Each phase of authorship represents a different form of interpretation of social structures and monastic strategy, which can be partly established by the individuals and places recorded in the miracle stories. The phases are key to understanding changes over time in the range and nature of the monastery's influence. Due to its exceptional length and detail, the collection lends itself to statistical analysis, and many of the individuals and properties mentioned in the collection are corroborated by other sources, including legal documents.[22]

Initial interest in using GIS for this data set stemmed from a desire to distinguish between authorial differences in the regions and types of land covered as well as the extent of the spatial landscapes that were invoked by each of the clerical authors. Geographically, the area covered is vast. Although most of the stories concern south-central France, locations in northern Spain, Italy and the Holy Land also receive mention. Besides their location, each place also has other attributes: a type (castle, church, urban area, etc.), the context in which the location is evoked (where a miracle occurred, or an individual came from, or a traveller visited), a reference in the source (i.e. Book I, Chap. 3) and a name (rendered in the original Latin or, as long as identification has been made, in its modern-day form). Despite some disruption following the French Revolution, geographic borders have remained relatively constant in much of southern France with many medieval counties roughly

[21]The most recent edition of the Latin text is that of Robertini (1994), hereon as *LMSF*. It is translated by Sheingorn (1995). On dating and authorship, see Robertini (1994).

[22]Those remaining for Sainte-Foy are mainly held in the Conques cartulary, cf. Desjardins (1879).

equivalent to their modern-day administrative successors.[23] Nevertheless, analysing zones of interest using textual descriptions alone is unsatisfactory, particularly due to the overlap in different units (dioceses and counties, for example) and the ambiguity and variation of some boundaries. The data also has a threefold chronological aspect: when a miracle supposedly occurred, when it was written down (as suggested by the book number) and, in relation to land claimed as property, sometimes a date when the property was donated or contested. Problematically, however, in the context of databases and the current capabilities of GIS technologies, dates are rarely precise—an issue that will be addressed later. Visualising the data using GIS enabled a range of cross-analyses of attribute types to be made, which transcended traditional avenues of enquiry.

The most significant revelation from the Sainte-Foy data set resulted from analysis of the locations of positive and negative miracles. The majority (73%) of the miracles of Sainte Foy are "positive" intercessions that aid a beneficiary, usually through healing, but also the liberation of prisoners, restoring lost property or survival from attack or accidental injury. Compared to the mean proportion of "negative" miracles (representing divine punishment) provided by Pierre-André Sigal's survey of the eleventh- and twelfth-century miracle stories in France, which stands at 12.6%, at 27% Foy's collection contains over twice the average (Sigal 1985, 291). Of these miracles, 14% describe the infliction of actual physical harm, and a further 10% resulted in death, usually suffered by those who had sinned against the monastery and its dependents. Comparing the geography of positive and negative miracles showed that the area in which the punishment miracles took place was significantly smaller than that covered by positive miracles. This substantiated other evidence to suggest that miracle types had differing objectives. Punishment miracles related directly to monastic social politics—they intended to interact with and influence local power brokers who had a direct bearing on the wealth and power of the monastery.[24] By narrating stories about saintly interaction with a person, either positive or negative, clerical hagiographers recorded and "sanctified" a relationship between the individual and the saint and therefore the monastery.

Mapping Sainte Foy's miracles thus provided evidence for the extent of Conques' self-constructed sociopolitical landscape, an area limited to south-central France but, significantly, given the administrative boundaries that often restrict such studies, extending much further than the county of the Rouergue alone. Whereas the written punishment miracles tend not to specify witnesses and often seem to have been creations of their monastic authors, affirmative miracles were frequently reported at the shrine by pilgrims and devotees giving thanks and were often testified to by multiple witnesses. Thus, the larger geographical area covered by locations related

[23] See the interview with Emmanuel Le Roy Ladurie in this book about the re-districting of France in the wake of the French Revolution.

[24] Foy's punishment miracles should be considered a "vengeance script" (White 2006, 172). Vengeance itself was part of normative feuding activity which was constructive in the building of power relations, as much as it could be destructive, for which the classic study is Miller (1990).

to positive miracles, reaching to Normandy, Italy, Catalonia, southern Spain and the Holy Land, represents the broader reach of the cultic influence of Sainte-Foy de Conques since tales of her beneficence were reported from far and wide. The disparity in the cultic and sociopolitical spheres of the monastery highlighted by visual representation of positive and negative miracles further reinforces the dual social and religious nature of the cult and the differing monastic comprehension and interaction with space in its interactions with lay society.

Cross-analyses of different attribute types like this are not dependent on GIS and could have been achieved using traditional cartographic means, although these are slower and less open to interrogation. The value of accessibility to mapping as a research tool by those new to the technology should not be underestimated, even for relatively small data sets and using the most basic of GIS capabilities. GIS effectively democratises the cartographic process, allowing autonomous historians to integrate geographic analyses into their research much more easily than before.[25] These analyses can demonstrate patterns and trends that the narratives themselves do not surrender so simply.

8.4 Wider Applications and Methodological Challenges

GIS not only offers opportunities to individual historians, it also allows studies to be cross-analysed more straightforwardly on a shared GIS platform. Combined with our understanding of major Roman and other medieval communication routes,[26] for example, the local roads listed for the Bobbio route could be used in conjunction with other sources and archaeological information, including the locations of *hospitalia* and other religious constructions that would have lined their sides, to reconstruct a picture of the medieval route map. In those isolated cases where we have more detailed information—such as for Archbishop Sigeric of Canterbury's journey to Rome just before the turn of the first millennium or the transport costs related to royal travel in the fourteenth-century England—we might construct probable routes and interesting diversions.[27] If this were integrated with chronological information including important political events and holy days, patterns in travel together with previously unnoticed connections might be made. This would become an extraordinary resource for historians studying any number of medieval journeys, from invading armies to merchants and pilgrims. GIS analyses of value to the individual researcher could also be integrated straightforwardly with online reference resources such as those described above.

[25] See also the chapter by Alexander von Lünen in this book for a discussion on this issue; Alexi Baker and Detlev Mares/Wolfgang Moschek also discuss this briefly in their respective chapters.

[26] See, for example, Margary (1973) and Talbert and Elliott (2010).

[27] See Ortenberg (1990) and Masschaele (1993). Other useful resources for this include Matthews (2007) and Renouard (1968).

Local studies in which territories and spheres of influence are constructed could also be more widely instructive if equivalent cartularies and miracle collections from other institutions were to be similarly modelled. We would have at our fingertips an extraordinary resource for testing spheres (perceived or claimed) of religious and political influence since cartularies—like miracle collections—were self-conscious constructions of communities.[28] Charter data for the monastic powerhouse of Cluny already exist in database form and have been analysed rigorously with significant conclusions being drawn regarding the social nature of land transactions (Rosenwein 1989). Since the giving and receiving of land was as much a social as an economic transaction, the resulting maps would also demonstrate networks of power relations. One function of GIS software of potential benefit is buffer zones to analyse areas of landlordship or jurisdiction (using charters of land transfers) and cultic influence (using miracles). Monastic charters need not be used only for the religious institutions themselves but could also—especially when cross-referenced with others—help to track areas of influence of particular families or networks. It may also provide an interesting way to approach historiography on kinship structures and the rooting of lineages to a particular territory ("topolineages" in French historiography).[29]

Despite the promise GIS offers on both individual and collaborative levels, significant methodological concerns present themselves. As well as general compatibility issues with software platforms, usually overcome by using web-based portals, and the multiplicity of projection systems, there are particular issues surrounding early medieval data. These include imprecise dates and uncertainty in location identification.

Representing uncertainty in the chronological aspect remains a central problem since change over time underpins all historical research, regardless of scale. Displaying this change in geovisual form is, for the amateur, often restricted to producing a series of static maps. Animation may improve the demonstration of chronological progression, enhancing the temporal characteristic of the data; likewise, a live map that could be interrogated for the situation at any given time. Presenting chronological development can be problematic since we often rely on a range of dates established by reference to the reign of a king, count, abbot or bishop, rather than specific and detailed chronological data, whilst on other occasions chronology is simply uncertain.[30] Visualising this uncertainty in a standardised form that is readily understandable presents another hurdle to be overcome before the chronological qualities of historical data can be utilised fully in map form.

[28] Bedos-Rezak (1994, 322). Mindful that these documents present methodological issues in their own right, due to the conscious and discriminate editing of their compilers and the process of copying, cartularies are nevertheless an extraordinary resource for a time period in which dense records do not survive (Cf. Guyotjeannin et al. (1993) and Kosto and Winroth (2002)).

[29] A process recently studied and re-dated by Le Jan (2000, 58).

[30] Recent computer technology has been used to help date medieval records, such as one strand of the Documents of Early England Data Set (DEEDS) project; see, for example, Spencer (2002).

The uncertainty of associating a place name to a GIS-compatible coordinate is perhaps the most significant barrier to capitalising on the potentially large resource of qualitative spatial data that might be mined from early medieval sources and, in particular, charter material. Converting spatial information in the form of a place name into a coordinate for use by GIS software—known as "georeferencing"—is a dangerous and complex process since these linguistic representations of human perceptions of space sit uncomfortably with the absolutism of visual mapping. Even if we accept the hazards of reductionism such an approach brings and attempt to identify each place as somewhere that can be mapped, we may still require place name specialists, local knowledge and historical cartography to overcome the problem of renaming, changing land use (including lost villages) and common names (for example, Châteauneuf, Villeneuve). This is in addition to the complexities of naming practices that often varied between Latin and vernacular usages. This limitation is not insurmountable but requires a committed manual input: there is no gazetteer for the medieval period that allows the more straightforward integration between text and map that modern GIS systems are able to utilise.[31] Neither is it likely that there will be one in the near future, if ever, that is detailed enough to be of use for thorough research since many place names appear only once in our sources.

For smaller data collections like those explored here, the problem is less acute since there is often a historiographical tradition of identification the historian will already be familiar with. Of 120 different locations in the Sainte-Foy miracles, for example, 16 remain unidentified and a further nine are disputed. On some occasions, there may be additional qualitative information in the source that allows us to identify roughly which region the place is situated in, but conflicting identifications can still be kilometres apart. In these cases, a decision must be made: either to discount the location entirely or to represent multiple alternative options. No standardisation yet exists for the latter option; thus, as it stands, maps need supporting explanatory text. On the other hand, GIS may offer some help towards the identification of locations, through the use of georeferenced historical maps, as will be discussed.

The cost analysis function is more complex than the cross-analyses suggested for the Sainte-Foy collection and presents more of a hurdle to medieval historians without GIS experience. Perhaps the biggest challenge to overcome before the function can be fully realised, however, is the lack of suitable georeferenced route information (in the form of vector layers) for a region in a given period. One way that we might conjecture about the presence and usage of local roads would be to use georeferenced historical maps. Significant headway is being made in this area by the David Rumsey collection, and such resources are increasingly accessible and searchable through Old Maps Online.[32] Vectorised major Roman routes have been made available by the *DARMC* and *RFO*, the former based on the *Barrington Atlas*

[31] On creating historical gazetteers, see Mostern and Johnson (2008) and Mostern (2008).

[32] http://www.davidrumsey.com, last accessed March 2012, see also Rumsey and Williams (2002, 16) and also http://www.oldmapsonline.org, last accessed March 2012.

and the latter on a variety of sources.[33] Impressively, accurate cartographic studies were compiled for much of Europe during the early modern era, for example, the Cassini series, produced during the seventeenth and eighteenth centuries by four generations of the Cassini family and covering much of Western Europe, especially France, or those of John Speed (1551/2–1629) and the road maps of John Ogilby (1600–1676) for England (Ogilby 1675). A georeferenced series of Cassini maps covering the whole of France is now available.[34] French historians frequently use the Cassini series as a fair guide to the landscape of pre-industrial and pre-Revolution France. Evidently, there will always be queries about the extent to which paths appeared or disappeared, or changed routes, and to which names were retained or changed. Conversely, the nature of topography, particularly in hilly regions like the Apennines, means that there has been infrequent and little change in roads or pathways since, given the natural features of the land, these are often the most obvious or only possible routes to take.

Georeferenced historical maps will also serve other uses, including helping to identify locations by recovering lost place names. In one example from the Sainte-Foy collection, the miracle stories tell us that the *villa* of Vialarels was "six miles from Conques" (LSMF, I.4). Historians have usually tentatively placed the villa in the commune of Firmi (44° 32′ 28″ N, 2° 18′ 42″ E), but plotting the data using a georeferenced Cassini base map suggested a more likely alternative at a nearby site marked as Vialarels, a toponym that has been lost since the map was drawn (44° 33′ 30″ N, 2° 14′ 15″; near the commune of Viviez). Plotting the sites on the base map narrowed down the search on these large, detailed maps and allowed this alternative identification to be made relatively quickly.

8.5 Conclusion

As with any database, it is the questions we ask of GIS that are important. The centrality of land to political, social and economic relations during the medieval period and the concomitant wealth of evidence concerned with physical locations give GIS an immediate relevance to early medieval history. Since investigations into the interaction of superstructures also rely on minute local studies, GIS offers opportunities at both levels. Once consensus is reached on a common coordinate system, GIS will facilitate analysis on a larger scale than has previously been possible via the integration of multiple data sets. For southern France alone, there are other collections of miracle stories and numerous cartularies like those composed at Conques which could similarly be visualised to demonstrate cultic,

[33] Talbert (2000); for a list of the latter's sources, see http://www.early-medieval-gis.blogspot.fr/, last accessed March 2012.

[34] http://demo.geogarage.com/cassini/ or http://webservices.cartosphere.com/, both last accessed March 2012.

jurisdictional or political spheres. If this data were combined, compared or layered, it could demonstrate the interaction of different religious institutions or public officials, outline administrative districts and reveal overlaps in spheres of control or interest, the acquisition and changing ownership of land. Equally, the possibilities for analysing routes like the small-scale example quoted for the San Colombano miracles could be applied on much wider scales to those of pilgrims, merchants or armies. When we consider other variables that could also be addressed, such as manuscript transmission, language and demographics, the use of historical GIS to facilitate understanding and to open new doors within medieval research is undeniable. Such grand ambitions would be most readily achieved with the collaboration of individual projects shared across a common web-based interface, given the minutely specific research that needs to be carried out at the micro-level.

There is no need for such ambitious scope to be disconcerting to the individual historian considering the use of GIS, and availability and cost is no longer an issue with various freeware options. Many historians already hold databases that could be converted into GIS-compatible data sets. The simplest functions of GIS, which are accessible to historians new to mapping software, more readily facilitate the consideration of the spatial dimension of narrative and other sources. As with the Sainte-Foy examples, spatial analysis of narratives can reveal patterns and trends that have previously passed unnoticed, leading to new research questions and hypotheses. Similarly, GIS may offer new propositions to debates such as that of bishop rivalry at Bobbio or help identify toponyms that were previously lost as at Vialarels. A general "upskilling" of historians across the board, and the resulting increase in the profile of the consideration of geographic aspects in research, will only serve to break down disciplinary boundaries further.

References

Armstrong J (2008) Local society and the defence of the English frontier in fifteenth-century Scotland: the war measures of 1482. Florilegium 25:127–149

Barthélemy D, White SD (1996) Debate: the 'feudal revolution': I and II. Past & Present 152:196–223

Bedos-Rezak B (1994) Diplomatic sources and medieval documentary practices: an essay in interpretative methodology. In: van Engen J (ed) The past and future of medieval studies, University of Notre Dame Press, Notre Dame (In.), pp 313–343

Bisson TN (1994) The feudal revolution. Past & Present 142:6–42

Bresslau H (1934) Miracula sancti Columbani. MGH Scr 30(II):993–1015

Davies W, Fouracre P (eds) (1995) Property and power in the early middle ages. Cambridge University Press, Cambridge

Desjardins G (ed) (1879) Cartulaire de l'abbaye de Conques en Rouergue. Picard, Paris

Devroey JP, Lauwers M (2007) L'espace des historiens medievistes: quelques remarques en guise de conclusion. In: Abbé JL (ed) Construction de l'espace au moyen age : pratiques et representations: XXXVIIe congrès de la SHMES: Mulhouse, 2–4 June 2006, Publications de la Sorbonne, Paris

Dreslerová D, Mikuláš R (2010) An early medieval symbol carved on a tree trunk: pathfinder or territorial marker? Antiquity 84:1067–1075

Duby G (1953) La société aux XIe et XIIIe siècles dans la région Mâconnaise. Armand Colin, Paris
Gregory IS, Ell PS (2007) Historical GIS: technologies, methodologies and scholarship. Cambridge Studies in Historical Geography, Cambridge University Press, Cambridge
Guyotjeannin O, Morelle L, Parisse M (eds) (1993) Les cartulaires: actes de la table ronde organisée par l'École Nationale des Chartes et le G.D.R. 121 du C.N.R.S. Ecoles des Chartes, Paris
Kosto A, Winroth A (eds) (2002) Charters, cartularies and archives: the preservation and transmission of documents in the Medieval West. Pontificial Institute of Mediaeval Studies, Toronto
Lauwers M (2005) Naissance du cimetière. Lieux sacrés et terre des morts dans l'Occident médiéval. Aubier, Paris
Layers E (2010) Turning toward place, space and time. In: Bodenhamer DJ, Corrigan J, Harris TM (eds) The spatial humanities: GIS and the future of humanities scholarship. Indiana University Press, Bloomington, pp 1–13
Le Jan R (2000) Continuity and change in the tenth-century nobility. In: Duggan A (ed) Nobles and nobility in medieval Europe: concepts, origins, transformations. Boydell, Woodbridge, pp 53–68
Lilley K (2009) City and cosmos: the medieval world in Urban form. Reaktion, London
Lilley K (2011a) Digital cartographies and medieval geographies. In: Daniels S, DeLyser D, Entrikin JN, Richardson D (eds) Envisioning landscapes, making worlds: geography and the humanities. Routledge, London/New York, pp 25–33
Lilley K (2011b) Geography's medieval history: a forgotten enterprise? Dialog Hum Geography 1(2):147–162
Lilley K, Lloyd C (2009) Mapping the realm: a new look at the Gough Map of Britain (c.1360). Imago Mundi 61(1):1–28
Lloyd C, Lilley K (2009) Cartographic veracity in medieval mapping: analyzing geographical variation in the Gough map of Great Britain. Ann Assoc Am Geograp 99(1):27–48
Margary I (1973) Roman roads in Britain, 3rd edn. Baker, London
Masschaele J (1993) Transport costs in medieval England. Econ Hist Rev 46:266–279
Matthews S (2007) Travel and travellers between England and Italy in the Anglo-Saxon centuries. Archaeopress, Oxford
Miller W (1990) Bloodtaking and peacemaking: feud, law, and society in Saga Iceland. University of Chicago Press, Chicago
Mostern R (2008) Historical gazetteers: an experiential perspective, with examples from Chinese history. Histor Method 41(1):39–46
Mostern R, Johnson I (2008) From named place to naming event: creating gazetteers for history. Int J Geogr Inform Sci 22(10):1091–1108
Nuvolone F (2000) Viaggiatori e pellegrini a Bobbio all'inizio ai *miracula Columbani*. Archivum Bobiense 3:73–119
Ogilby J (1675) Britannia volume the first. Or, An illustration of the kingdom of England and dominion of Wales: by a geographical and historical description of the principal roads thereof. Printed by the author, London
Ortenberg V (1990) Archbishop Sigeric's journey to Rome in 990. Anglo-Saxon England 19:197–246
Renouard Y (1968) Etudes d'histoire medieval, vol 2. S.E.V.P.E.N., Paris
Reuter T, Wickham C (1997) Debate: the 'feudal revolution', III and IV. Past & Present 155:177–208
Robertini L (1994) Liber miraculorum sancte fidis. Centro italiano di studi sull'alto Medioevo, Spoleto
Rollason D (1985) The miracles of St. Benedict: a window on early medieval France. In: Mayr-Harting H, Moore R (eds) Studies in medieval history presented to R.H.C. Davis. Hambledon, London, pp 73–90

Rosenwein B (1989) To be the neighbor of Saint Peter: the social meaning of Cluny's property, 909–1049. Cornell University Press, Ithaca
Rosenwein BH, Head T, Farmer SA (1991) Monks and their enemies: a comparative approach. Speculum 66:764–796
Rumsey D, Williams M (2002) Historical maps in GIS. In: Knowles AK (ed) Past time, past place. GIS for History, ESRI, Redlands, pp 1–18
Sheingorn P (1995) Book of Sainte Foy. University of Pennsylvania Press, Philadelphia
Sigal PA (1985) L'homme et le miracle dans la France médiévale: XIe-XIIe siècle. Editions du Cerf, Paris
Spencer A (2002) Dating charters using textual evidence. In: Keats-Rohan KS (ed) Resourcing sources. The use of computers in developing prosopographical methodology, prosopographica et genealogica, vol 7. Unit for Prosopographical Research, Linacre College/University of Oxford, Oxford, pp 193–200
Talbert R (2000) Barrington atlas of the Greek and Roman world. Princeton University Press, Princeton
Talbert RJA, Elliott T (2010) Rome's world: the Peutinger map reconsidered. Cambridge University Press, Cambridge
Taylor F (2012) Miracula, Saints' cults and socio-political landscapes: Bobbio, Conques and post-Carolingian society. Unpublished PhD thesis, University of Nottingham
Theuws F, Jong MD, van Rhijn C (eds) (2001) Topographies of power in the Early Middle Ages. Brill, Leiden
Tosi M (1981) Il trasferimento di S. Colombano da Bobbio a Pavia : 17–30 luglio [929]. Archivum Bobiense 3:129–150
Ward B (1987) Miracles and the medieval mind: theory, record and event 1000–1215, Rev. edn. Wildwood House, Aldershot
White S (2006) Garsinde v. Sainte Foy: argumentation, threat and vengeance in eleventh-century monastic litigation. In: Jamroziak E, Burton JE (eds) Religious and laity in western Europe, 1000–1400: interaction, negotiation, and power. Europa sacra, vol 2. Brepols, Turnhout, pp 169–182
Wickham C (1994) Land and power: studies in Italian and European social history, 400–1200. British School at Rome, London

Chapter 9
The Role of Knowledge Spaces in Geographically-Oriented History

Monica Wachowicz and J.B. Owens

9.1 Introduction

Space matters, not for "the simplistic and overly used reason that everything happens in space, but because where things happen is critical to knowing how and why they happen" (Warf and Arias 2009, 1). The term spatial history was firstly coined in the 1980s by Paul Carter in *The Road to Botany Bay: An Essay in Spatial History* (Knowles 2005, 12). The theoretical foundations that emerged from this spatial turn in history have exposed the recognition that place, space and time are centrally and inevitably involved in the design of historical GIS (Bodenhamer 2010; Goodchild and Janelle 2010). History of any place is shaped by the way people relate to this place, which is connected to other places and the events that change these connections over time. Place has to be one of the most multilayered and multipurpose words in our language (Harvey 2003[1969]). Their origins, meanings, use and typology can vary within and between historical periods, which may cover small to large spatial extensions, and which sometimes become unstable, leading to the emergence of social networks, evolution and chaos.

Another aspect of spatial history is the issue of transformation of historical data from their original representation to alternative spatial representations. Spatialisation refers to the construction of abstract spaces of knowledge that can aid in visualisation, pattern detection and the accumulation of scientific insight (Goodchild et al. 1993; Skupin and Fabrikant 2003). Therefore, things that are not explicitly spatial (e.g. social and kinship networks) as well as text documents (e.g. newspapers,

M. Wachowicz (✉)
Geodesy and Geomatics Engineering department, University of New Brunswick, New Brunswick, Canada
e-mail: monicaw@unb.ca

J.B. Owens
History Department, Idaho State University, Pocatello, ID, USA
e-mail: owenjack@isu.edu

A. von Lünen and C. Travis (eds.), *History and GIS: Epistemologies, Considerations and Reflections*, DOI 10.1007/978-94-007-5009-8_9,

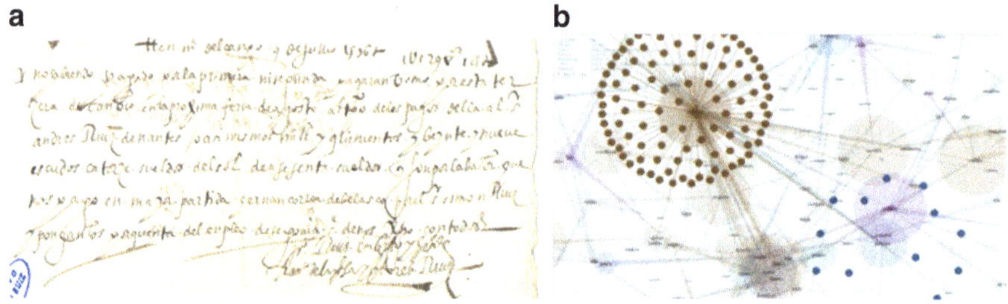

Fig. 9.1 The spatialisation of the commercial letters obtained from the Simón Ruiz's archive. (**a**) Example of the unstructured text data from the 200 letters corresponding to the years 1560–1580 (**b**) The spatial representation simulating a 2D particle system. Each particle represents a word and its position in relation to the relative position this word occupies in the text of a letter (© Monica Wachowicz)

reports and letters) can be represented using spatial representations. For example, spatialisation can be used to help historians enhance their intuition through the visual exploration of unstructured data found in historical texts such as the set of Simón Ruiz's sixteenth-century commercial letters. In these letters, information about personal nature (e.g. "this is a sorry land for trading") is intertwined with more objective data (e.g. "the wheat was sold at 16 ducats"). This information is further tangled with opinions, memories and expectations of a personal nature, which in turn may be linked to geographical places as well as social and trade-related spaces. Figure 9.1 illustrates how the creation of a text-mining system capable of automatically reading the letters and identifying keywords can be transformed into a particle visual representation for those words found during the reading.[1]

However, Warf and Arias (2009) argue that modernity has overly represented space as a series of spatial layers which are usually populated by homogeneous objects or fields. This is definitely the situation in current geographical information systems (GIS) where layers are used to structure, integrate, manipulate, analyse and display georeferenced data sets (i.e. structured data) rather than text and image data sets (i.e. unstructured data). Nevertheless, GIS has proven to be useful to structure, analyse and visualise historical data, in ways that were either completely new or were made significantly easier for historians (Gregory and Ell 2007; McIntosh et al. 2011).

The advent of large data repositories in digital humanities (e.g. The Archaeology Data Service, Early Canadiana Online and Opening History) has opened an unprecedented digital access to materials used by historians, ranging from digitised books, newspapers, historical maps, registers and letters. In view of that, Goodchild (2008) points to the expected shift that will change the current single disciplinary view towards the realms of integrated spatial knowledge, which will arise from the semantic notions of space where things (e.g. networks, places and feelings) are represented based on partial, incomplete, never fully known information. This shift

[1] More details about the system can be found in Iturrioz et al. (2009).

is fundamental for social sciences and humanities, involving a rethinking of the very notion and significance of spatialisation in terms of how space itself is theorised, what it means and how it is made.

Towards this end, Owens (2007, 2010) has coined the term "geographically-integrated history" to explain a transformative research and interdisciplinary teaching strategy, founded on the conceptualisation that (a) the understanding of historical events requires an integration of the natural, social and cultural environments on the basis of place, space and time and (b) accomplishing this integration poses a challenge that can be met with modern computational tools, especially dynamic forms of GIS and social network analysis, and visualisation techniques.

In this chapter, we propose a knowledge space representation for helping construct formal and linguistic knowledge in geographically integrated history. From a modelling perspective, a knowledge space representation can provide an important insight into geographically-integrated history by identifying concepts as symbolic forms that describe the logic underlying thought and reasoning of abstract assumptions made by historians. From a philosophical point of view, knowledge spaces are significant as they organise abstract theories within a systematic and axiomatic framework. From a computational perspective, a knowledge space representation can provide a simple and computationally efficient mechanism for facilitating knowledge reuse.

9.2 What Is a Knowledge Space?

Knowledge spaces are founded on the Cassirer's philosophy of making imaginable spaces that are filled with symbolic forms representing the world (Cassirer 1923–1929). They can be seen as "a strong pair of glasses that determine what we can see bringing some part of the world into sharp focus at the expense of blurring other parts" (Davis et al. 1993, 19). This focussing/blurring metaphor is useful for grasping the fundamental role of knowledge spaces in observing, interpreting and understanding the world (Fig. 9.2). Knowledge spaces are an answer to the question: In what terms should a historian think about the world?

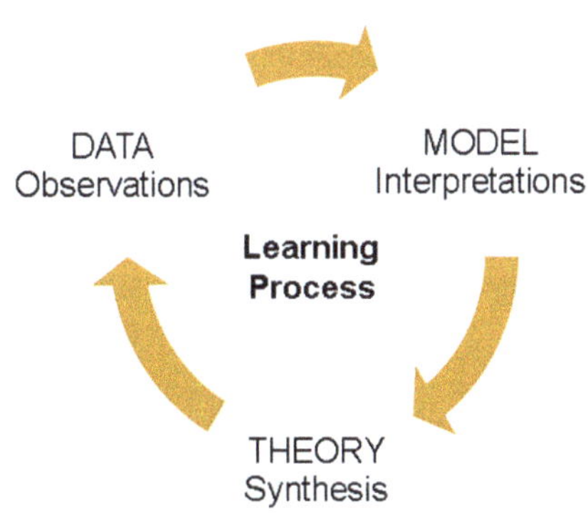

Fig. 9.2 Cassirer's imaginable spaces in a learning process (© Monica Wachowicz)

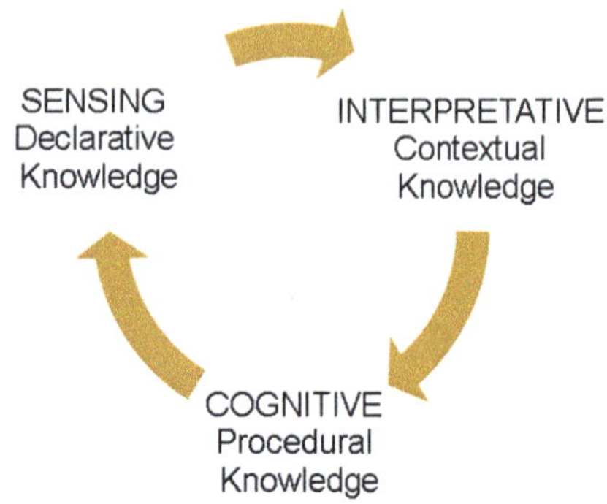

Fig. 9.3 The three knowledge spaces (© Monica Wachowicz)

A knowledge space is a representation of some logically coherent situation or potential reality, where various symbolic forms are treated as true, they are assumed to exist, and relations between them are supposed to hold at a specific place, space and time, having certain existential assumptions that express the historians' generalisations and beliefs. The notion of symbolic forms is based on the concept of a knowledge level model (Newell 1982). Symbolic forms may be atoms in physics, molecules or cells in biology or networks in socio-economics. When considering how many symbolic forms are needed to be represented in a knowledge space, we should make a clear distinction between complex and complicated knowledge spaces (Cilliers 1998). A complicated knowledge space is one with a large number of symbolic forms. For example, a social network portraying a systemic behaviour can be sufficiently represented by its symbolic forms of which it is made, and to a manageable extent, by modelling their connections over time. In contrast, a social network in a complex knowledge space is one whose behaviour cannot be adequately understood by only representing its various components and connections, but instead, by unveiling the global patterns generated from the local interactions among its components. Therefore, we should never only represent isolated symbolic forms and their connections but also model their dynamic interactions over place, space and time (Oyama 2000).

Three knowledge spaces are presented in this chapter for supporting a progressive learning process of symbolic forms in geographically-integrated history (Fig. 9.3). First, the sensing knowledge space begins with the observation of places, space and time based on declarative knowledge, usually expressed in propositional sentences or indicative geographical features, events, facts and persons' descriptions. The historians' notion of what is random and what is complex in the world depends directly on the subjectivity of their observations. Since historical data is typically text-heavy, imprecise and vague, the inconsistency and uncertainties in any observation can lead to critical consequences for the usability of a historical GIS.[2] The sensing knowledge space plays an important role in helping historians not only to define "what is observed?" but also "how does one observe the past?" Therefore, the

[2]For more details in usability issues, see Hunter et al. (2002).

symbolic forms represent how historical data will be combined to produce results that satisfy the information needs. In particular, what information principles need to be represented in order to explain, at least in a pragmatic manner, a vast variety of historical data which are already available or are found under the database structure of digital repositories. The design of a sensing knowledge space is about relating hitherto unstructured and structured data and making them intelligible inside a comprehensive and coherent representation.

However, "meaning" does not adhere to the symbolic forms themselves but relates directly to what is asserted by the use of them and what other uses, available in the sensing knowledge space, are implicitly rejected. Therefore, the interpretative knowledge space is required for constructing the image representation that is needed for signifying higher and more complex social and cultural forms, which assume additional significance when interpreted within a context. For example, in economics, it has been natural to extend Shannon's classic work on information theory to develop stochastic simulations in order to understand how different variables, such as the number of producers and the market size of a commodity, could be understood in terms of more complex concepts such as scarcity, growth and depression (Chen 2005). This was made possible under the assumption that all human activities represent extraction and transformation of low entropy from a system, making it natural to relate economic variables to low entropy.

The interpretative knowledge space is the kind of knowledge representation most dependent on past experience for interpretation and most frequently prone to misinterpretation due to contradicting information sources as well as opposing historians' generalisations and beliefs. A "context" is a situation in place and time at which one or more historians act, including by speaking or writing, to assert a "meaning" that involves an interpretation of some observations of this situation. Contexts are "local models" that represent a historian's view about the inferences. In contrast, ontologies[3] are "shared models" of some domain that encode a view which is common to a set of different parties. The context complexity stems from the failure to consider the role of an interpretative knowledge space in the shaping of meaning by defining symbolic forms where contradictions can be detected and made explicit and their coherence can be questioned. The issue of determining the meaning and importance of linguistic elements used in a particular context over space and time has too often been pushed to the background when it must be central in the design of a historical GIS.

Third, the cognitive knowledge space underlies a diverse array of communication-related issues associated with message production, and in particular, the analysis of cultural influences on communication and collaboration (e.g. Applegate 1990; Kelly 1955). Historians are interested in communicating their understanding of a

[3]An ontology is defined here after Gruber (1992, 1): "[A]n ontology is a description (like a formal specification of a program) of the concepts and relationships that can exist for an agent or a community of agents. This definition is consistent with the usage of ontology as set-of-concept-definitions, but more general. And it is certainly a different sense of the word than its use in philosophy. What is important is what an ontology is *for*".

Table 9.1 Main characteristics of knowledge spaces

Sensing knowledge space	Interpretative knowledge space	Cognitive knowledge space
Propositional representation	Image representation	Procedural representation
Observation complexity	Context complexity	Communication complexity
Explanation	Exploration	Collaboration
Element encoding	Pattern discovery	Principle generation

historical path from "context A" to "context B", leading to the construction of procedural knowledge on how to relate dynamic patterns to high-level properties for a given context. The symbolic forms with differentiated, articulated and integrated linguistic elements will have greater information communication capacity in a cognitive knowledge space, which in turn will facilitate the representation of the communication-relevant symbolic forms, and thus assist the production of historian-centred messages that will reflect an awareness of and an adaptation to the subjective, affective and relational aspects of geographically-integrated history. However, symbolic forms here should not be seen as a personality characteristics' representation of a historian's point of view but, on the contrary, it should be seen as a communication learning process.

Table 9.1 provides an overview of the main characteristics of these knowledge spaces.

9.3 Sensing Knowledge Space

The sensing knowledge space encapsulates the historians' declarative knowledge, meaning the symbolic forms for representing the complex toponyms used for naming places, space and time, as well as the dynamic connections among them. Golledge and Stimson (1987) point out that declarative knowledge is at a lower level of a learning process which is based on defining symbolic forms as "packets of information that contains variables" (Rumelhart and Norman 1985). The scientific premise here is that a learning process is needed for selecting a set of symbolic forms that will offer significant insight in the development of systematic approaches for gathering observations in a historical GIS.

Historians are already facing the obstacles imposed by recovering, assembling and systematising very heterogeneous texts and turning them into structured data that can be stored as fields and records into a database. One prominent aspect of observing complexity being currently dealt with by historians is the one originated by georeferencing the diversity among the names of geographic features found in historical data (Southall 2003). The inconsistency arises from the toponymy changes in historical data during chronologically extended periods over time that can lead to overlapped, unlinked or unrelated georeferences in historical GIS. The GeoNames geographical database covers all countries and contains over eight million place names which are usually employed for georeferencing historical references, but unfortunately, it is not appropriate for the categorisation used in historical data sources. For example, although Pedro Juan Villuga's 1546 guidebook talks about "Spanish" commercial and pilgrimage routes, "Spain", in the absence of a country of that name, meant the entire Iberian Peninsula (Villuga 1546). Should we keep the past location/extension of Spain? Or convert to the present location/extension of the country as it is today? How to avoid inaccurate/misguided information about the location of Spanish routes may be communicated to an inexperienced historical GIS user?

More research is needed in order to represent the scaling and evolutionary aspects of toponymy of geographical features in a historical GIS. The implications of such sometimes imperceptible nuance might also compromise the assignment of data to a geographical feature and, in consequence, compromise the architecture of a database and further map reading and analyses. This is not a simple task, and as Polónia et al. (2010, 113) properly pointed out, the "choice between the past or present territorial locations/extensions of geofeatures, or even the idiom that should be used to identify those features, will always have to be discussed and agreed between multiple and multidisciplinary teams." The success of a historical GIS relies on developing a systematic approach that considers and describes those choices. We propose the development of symbolic forms to achieve that by identifying a set of explanatory concepts that will generate patterns of interest. Some symbolic forms have already been proposed in the literature. The approaches have been developed on adhoc basis, and they are usually focussed on supporting database requirements.

One example is the TimeLink database[4] which was designed as a web application for allowing access to the bills of exchange of Simón Ruiz, a New-Christian merchant from Medina del Campo, Castile, Spain. The actor-based symbolic form was used for gathering the relevant attributes for the reconstitution of "biographies" of merchants and places using linear time. This symbolic form was implemented as a set of classes that have attributes related to the identification of individuals, their personal information (including their marital status, family member relationship, profession and religion) and their societal functions (including economical, social, political and family lineage). In contrast, the TimeMap software[5] has used time filters to allow the animated maps display of historical events which were stored in a collaborative social bibliographic database named as Heurist.[6]

We propose the design of seven symbolic forms in the sensing knowledge space, (Fig. 9.4).

Time. It represents the temporal dimension which can be represented as linear and cyclic time (Frank 1998). Linear time dimension is a continuum in which time can be considered as having a time point and duration. A time point might be a date, hour, minutes when some event or activity has taken place. Duration is

[4]http://www.iclio.pt/, last accessed 27 Oct 2011.

[5]http://www.timemap.net/, last accessed 27 October 2011.

[6]http://www.HeuristScholar.org, last accessed 27 October 2011.

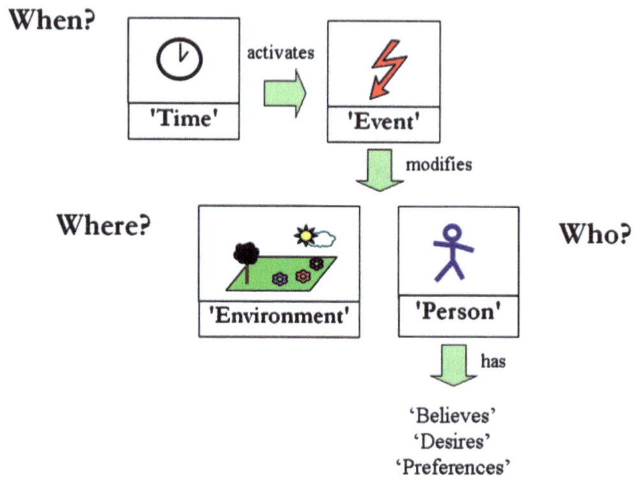

Fig. 9.4 The symbolic forms and respective relations in the sensing knowledge space (© Monica Wachowicz)

defined as an interval or period of time between two points in time. On the other hand, due to the similar repeating patterns over time, time can also considered as cyclic (Hornsby et al. 1999). In this case, the behaviour may change with periodic regularity, at different temporal scales (e.g. seasons of year, days of week, moment of the day). For example, in Britain during the later Middle Ages, pneumonia undoubtedly occurred in epidemic form in the winter months (Baker 1976).

Event. An event is an action that occurs in time and changes the characteristics of the features of a physical, social and/or economical environment as well as the preferences, desires and beliefs of a person. Events are very important in the sensing knowledge space as they explain the main causes of observed changes (Langran 1993; Kaneiwal et al. 2007; Zhang 2005). Some examples of events may include (a) weather-related events such as rain and snow acting over a physical environment to as diverse as earthquakes acting over an economical environment such as a financial market; (b) man-made-related events such as the "tulip mania" in the Netherlands that wilted suddenly in 1637 to the South Sea Bubble that ended with the first huge market crash in England in 1720; and (c) activity-related events such as the flow of visitors, patients and staff play in preventing disease transmission in healthcare facilities.

Environment. It is the physical, social or economical environment where events occur.

Person. It represents the agent, whether it is a person or a group of persons sharing the same beliefs, preferences and desires, usually based on a common goal.

Preferences. It represents the preferences of a person represented in the sensing knowledge space. Motivation-based typologies are frequently used for defining preferences as well as not scientifically founded description based upon historians' statements and expressions.

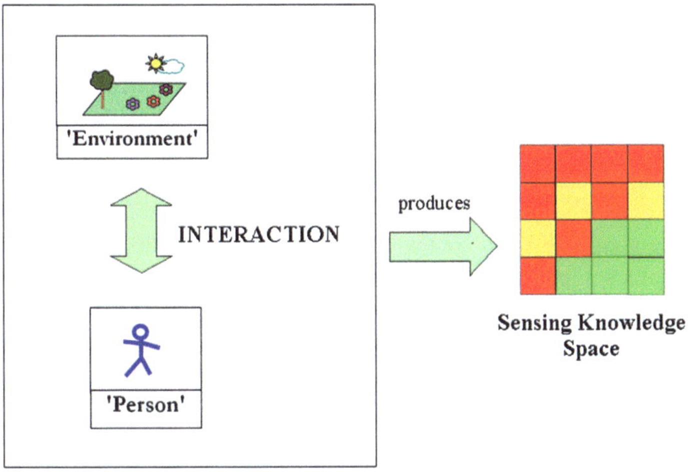

Fig. 9.5 An interaction between the environment and the person forms can produce a spatial representation (e.g. grid cell) of a sensing knowledge space (© Monica Wachowicz)

Beliefs and Desires. Persons generate beliefs about the current state of an environment based on their perceptions of it. They observe the environment and acquire information about its aspects related to a specific desire. This results in a set of beliefs accounting for the current state of the environment according to an individual person.[7]

It is through the interaction between the symbolic forms that a sensing knowledge space is produced (Fig. 9.5).

In the sensing knowledge space, visual analytics is important for achieving synthesis and presentation. Map-based techniques can allow the historical data and its visual appearance to be changed interactively (Dykes 1997). Map legends are often used as the basis for interaction, as shown by Peterson (1999) and Andrienko and Andrienko (1999), permitting the historians to change the appearance of the elements mapped and thereby define and possibly explain emergent patterns such as clusters, relationships and changes over space and time. Projection techniques can also be used for statistical transformations such as principle component analysis and multidimensional scaling in order to project structure or trends from the data (e.g. Asimov 1985; Haslett et al. 1991; Cook et al. 1995). They are also often based around scatter plots. Finally, hierarchical and network techniques can help historians to organise their historical data according to a specific data structure, such as a tree (Robertson 1991) or network (Huffaker et al. 1999), with progressive levels refining the display into subspaces.

[7]See Wooldridge (2000) for a detailed description of BDI models in agent-based modeling, which characterises agents using anthropomorphic notions, such as mental states and actions.

9.4 Interpretative Knowledge Space

After systematising their approach for gathering their historical data in historical GIS, historians are coming to realise the need for supporting a learning process as well (Owens 2010; Southall et al. 2009). This process plays a role in understanding how historians contextualise their universe of discourse. The interpretative knowledge space aims at supporting such a learning process by an image representation, whereas the goal is to build contexts that can function in the absence of predetermined hypotheses, especially if little is known concerning structures in data (e.g. when searching for a hypothesis to test analytically). Lackoff (1987) suggests that image representation provides a format for encoding information from vision and language simultaneously. This may be one explanation for the early development of maps in the history of civilisation and the important role they have played (and continue to play) in attempts to understand reality. For example, networks and connections between nodes need an image representation in order to be wholly understood. In this case, maps are needed when a historian would like to understand collective patterns and relationships because we simply cannot separate patterns from context (Kraak and Ormeling 2003).

A context is defined here as a psychological construct (Masyn et al. 2010), a subset of a historian's assumption about the world that plays a determinant role on the exploration of historical data. This makes quite difficult to find an agreement on what logical structure is suitable when context dependent information is involved. Therefore, it is when inductive reasoning becomes useful since it is arguably the most reliable means of learning due to the availability of many robust, automated methods developed by the machine learning community.[8]

In the interpretative knowledge space, the task becomes one of uncovering related hypotheses to explain the discovered patterns. Consequently, induction is most commonly applied after category construction and training examples have been developed in the sensing knowledge space (Fig. 9.6). Some examples of tasks include prediction, classification, clustering and association. Different methods can be used to perform these tasks, such as statistical methods, case-based reasoning, neural networks, decision trees, rule induction, Bayesian belief networks, genetic algorithms, fuzzy and rough sets theory.

An important aspect that cartographers are beginning to grasp is the choice of context representation. For example, they have usually assumed that there is only a unique way of categorising any data set by relying on quantifiable variables of categories to depicting visual representation. There is a need to explore the possibility of varying levels of contextualisation for different goals, applications and perspectives and explore how visual analytics might incorporate some of the less precisely defined (but no less truthful) ways of categorising the world. The argument made by Harley (1989) for "deconstructing the map" is that cartographers have (perhaps unconsciously) conspired to present maps as a direct window on the world rather than a highly processed representation of selected contexts of it.

[8]See Mitchell (1997) for many examples of inductive tools.

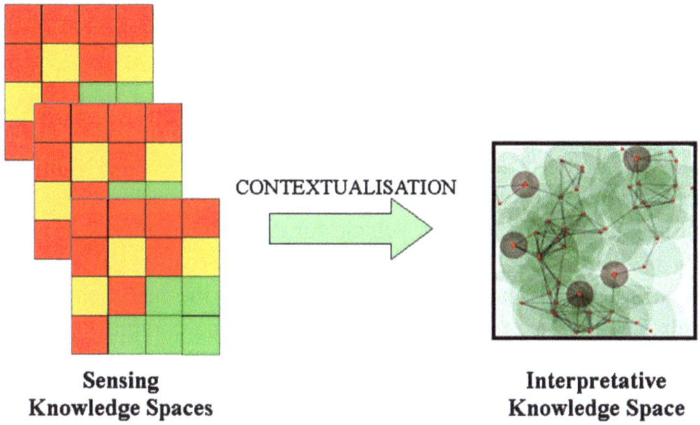

Fig. 9.6 The contextualisation of sensing knowledge spaces into an interpretative knowledge space (© Monica Wachowicz)

Visual analytics plays an important role in the exploration of contexts in the interpretative knowledge space. Chart-based techniques can plot the data on a chart or graph, common examples being scatter plots and parallel coordinate plots. Scatter plots use a simple 2D or 3D graph with dots or spheres to mark the position of individual data items (Cleveland and McGill 1988). Parallel coordinate plots employ a (usually larger) number of parallel axes through which a trace of each data item can be made (Inselberg 1985). These techniques are often accompanied by linking and brushing methods, allowing selected data points to be viewed in different ways or within different axes (e.g. Buja et al. 1996). Pixel techniques map data values to individual pixels that are ordered on the screen so that datastreams of similar values produce visible clusters in 2D (Jerding and Stasko 1998). The screen can be divided into separate windows if several attributes are to be visualised (Keim and Kriegel 1994). Finally, iconographic techniques use complex symbols, such as stick figures (Pickett and Grinstein 1988) or faces (Chernoff 1973; Dorling 1994) to encode many data dimensions simultaneously. The aim of iconographic displays is to promote perception of the "whole" while still allowing some differentiation of individual variables. A highly interactive visual environment might prove to be a very intuitive setting for defining training examples from which to construct contexts, for example, or otherwise explaining contexts definitions (Inselberg and Avidan 1999).

Some examples include the use of Hotelviz, an integrated software for visual data analysis, to help extract the complex visitation patterns that are to be found in historic hotel guest registers of commercial hotels in three small places in Central Pennsylvania during the late nineteenth century (Fyfe et al. 2009); and the work of historian Dan Edelstein on digitally mapping the Republic of Letters in order to visually reconstruct the network of Enlightenment correspondence that linked thinkers like Locke, Voltaire and Rousseau (Cohen 2010).

9.5 Cognitive Knowledge Space

The cognitive knowledge space can be approached as the last step in the learning process that has inputs, transmission and reception of information. It is concerned with the dissemination of and access to through communication symbolic forms. Communication should be viewed as the primary function of a historical GIS and the symbolic forms in this case are the vehicle for that communication. The cognitive complexity is identified by the numerous obstacles or filters that historical data must pass through on its route from reality through the historian understanding of it. As already proposed by MacEachren (1995), the communication symbolic forms integrate the cognitive-semiotic variables that will act as filters that include objectives, prior knowledge and experience, perceptual and spatial abilities, understanding of symbol system, external considerations (e.g. map readers demands), preconceptions and the abstraction process by which information is put into a map form (e.g. projection, simplification, generalisation, classification, symbolisation). According to communication theory, each of these symbolic forms can act to inhibit information transmission, resulting in information loss or uncertainty in communication. The research challenge is to develop multifaceted, multilevel, multi-perspective representational approaches of rhetorical discourse in which historians can structure their cognitive knowledge and the ways in which symbolic forms applied to the resulting representations restructure that knowledge and yield multiple alternative representations.

Visual analytics might bring two advantages to cognitive knowledge spaces. The first is the high degree of interaction, allowing a historian to explore data from several perspectives and steer the learning process (e.g. Ankerst et al. 1999). The second is the rich visual environment, allowing great variability of data to be viewed concurrently. Within such a setting, one might envision communication forms that could be used to *explain* the findings that the historian has encountered. Consequently, the need to combine maps with subjective information emerges naturally as shown in projects such as The Atlas of Emotions (Bruno 2002), the Atlas of Literature (Piatti et al. 2009) or the Atlas of Cinema (Caquard 2008). People are usually linked to specific locations that they know very well, and towards which they have developed affection or emotions, both positive and negative. If those people already have the chance to talk about those places employing modern collaboration tools and to supply information on places they know well for the collaborative creation of maps, it is impossible not to conceive a new, emerging cartography in which information is of a personal nature and whose subjective quality gives place to what Cartwright and his collaborators have coined as affective cartography (Cartwright et al. 2008). It is possible to differentiate between two types of affective mapping depending on whether the collaboration of individuals is required or not (Iturrioz and Wachowicz 2010). When affective mapping is carried out by means of collaboration, the starting point for the creation of maps is often of a subjective nature. In this case, the historian asks the participants for a series of data, collects a list of opinions about the dwelled space (be it known or remembered), then processes and codifies them and finally gives shape to new visual representations. In another category, we would place those affective maps made by only one individual or agent. In these cases, the affective aspects stem from the capacity to stir up responses in the map's receiver, be it due to the subject of study or to any other particular treatment (for example, the way it is shown).

In terms of reasoning, one goal of the cognitive knowledge space is to build tools that can utilise an abductive form of reasoning. In the cognitive knowledge space, abductive reasoning is aimed at discovering patterns within the data while simultaneously proposing a hypothesis by which the patterns might have come to be. In computational data mining, the hypothesis is usually constructed as some kind of classification scheme or set of association rules (e.g. Han 2002) by which useful structure can be imposed on the data. In some forms of visual exploration, an abductive task is performed collaboratively between the historian and the visualisation; patterns are observed as a consequence of how scene is constructed and rendered and how the historian perceives and comprehends it. The simultaneous task of hypothesis generation is also similarly split; the mappings used to create the scene themselves convey a hypothesis, and a historian may generate one or more internal theories to explain the observed structure. Abductive methods may seem more appealing, but they can also be less reliable because the visual patterns produced may be irrelevant, anomalous or misleading, and there is no way of guaranteeing their statistical significance, since that would presuppose the existence of a model (i.e. the sensing knowledge space).[9]

9.6 Research Challenges

This section considers the research challenges arising in developing knowledge spaces. The means to achieve effective knowledge acquisition, construction and dissemination in geographically-integrated history is still largely an open question. Many of the challenges described here are shared across a number of scientific disciplines, but others are unique to a geographically-integrated history setting, and relate either to the unique properties of historical information or to the three types of knowledge spaces described in this chapter. Therefore, the challenges have been posed as research questions and framed in terms of the different knowledge spaces.

Sensing Knowledge Space: What are the different tasks that should be involved in gathering historical data? To what extent do the different georeferencing techniques affect logical inference? What kinds of symbolic forms work best for particular tasks, applications or historical data? How to ensure that observed structure in the historical data must be related to a supporting hypothesis of some sort? How does the presence or absence of different historical data sets affect the usability of a historical GIS? What kind of computational architectures are

[9]See Alexander von Lünen's chapter in this book for a very extensive discussion of abductive reasoning in history and GIS.

most efficient and effective for integrating digital repositories with a historical GIS? When should we measure more, and when should we represent better in the sensing knowledge space? How is the uncertainty in the knowledge representation related to uncertainties in the input historical data formulation?

Interpretative Knowledge Space: How should the interpretative knowledge space take advantage of newer developments in visual analytics and languages information visualisation? What is the role of context in presenting the a priori knowledge of a historian in such a way that it will help the understanding of the symbolic forms used in an interpretative knowledge space? Can a taxonomy of operations based on context be developed to relate tasks historians are trying to accomplish with appropriate tools and functionalities of a historical GIS? When is it possible to average small scales into large-scale effects without affecting the inferences made within an interpretative knowledge space?

Cognitive Knowledge Space: How should findings be communicated in a cognitive knowledge space? In what way do various historians, communication and collaboration strategies affect the design of a cognitive knowledge space? Do historians working collaboratively have more or less success in knowledge dissemination, and how does their pooled expertise affect the design of a historical GIS? What is necessary to encourage collaboration in building a cognitive knowledge space based on a discipline characterised by solitary research and publication?

9.7 Conclusions

As the previous sections show, the field of geographically-integrated history embraces many different research questions from the perspective of the data, the system, visual techniques, modes of inference and collaboration. In particular, the design of historical GIS is about relating hitherto unstructured data and making them intelligible inside a comprehensive and coherent representation (i.e. model). This is only going to be possible if a historical GIS can support different levels of knowledge by considering different representations of how historians can observe, interpret and communicate a specific event, network, situation, pattern and principle. Three knowledge spaces have been proposed for taking on the observing, context and cognitive complexity levels into a logical representation of some situation or potential reality, where various symbolic forms are treated as true, they are assumed to exist, and relations between them are supposed to hold at specific places, spaces and times (Fig. 9.7).

The knowledge spaces shows the benefits of shifting from the strict quantitative approach of GIS towards qualitative explorations of unstructured data by bringing together the fields of knowledge representation, data mining, natural language processing, machine learning, information retrieval and visual analytics. This chapter proposes three knowledge spaces for building symbolic forms that represent the declarative, image and procedural knowledge about geographically-integrated history as a complex and dynamic representation, which is spatially large, and

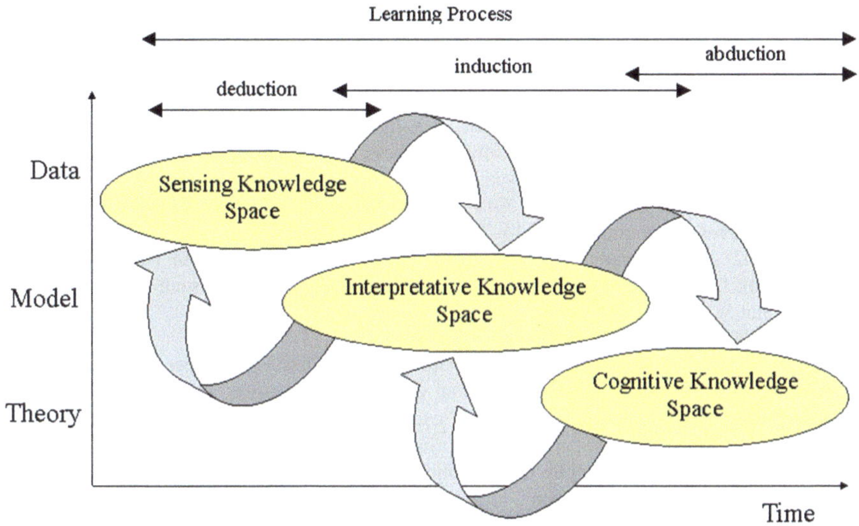

Fig. 9.7 Different knowledge spaces in the learning process and the modes of scientific inference that apply to each space (© Monica Wachowicz)

in more recent centuries, global in extension, and which sometimes becomes unstable, leading to a phase transition, bifurcation and the organisation of new representations. Within a knowledge space, place, space and time are connected by symbolic forms, which are the sources of innovation and the emergence of new forms over space and time.

Finally, this chapter has identified promising research topics that might extend, challenge or reaffirm our prevailing, quite often implicit assumptions on the design of historical GIS. At the very least, this chapter has clarified those areas of cross-disciplinary exchange that we might decide to explore further.

Acknowledgements This research was carried under the SOCNET project (understanding social networks within complex, nonlinear systems: geographically integrated history and dynamics GIS) that is a multidisciplinary and multinational research project funded by the US National Science Foundation (NSF) Office of Cyberinfrastructure (OCI), 2009–2013. I would like to thank Prof. Amélia Pólonia and Sara Pinto (University of Porto, Portugal) for our fruitful discussions on the research challenges in the field of geographically-integrated history.

References

Andrienko GL, Andrienko NV (1999) Interactive maps for visual data exploration. Int J Geogr Inf Sci 13(4):355–374

Ankerst M, Cand Ester EM, Kriegel HP (1999) Visual classification: an interactive approach to decision tree construction. In: KDD'99 proceedings of fifth ACM SIGKDD international conference on knowledge discovery and data mining, San Diego, pp 392–396

Applegate JL (1990) Constructs and communication: a pragmatic integration. In: Neimeyer G (ed) Advances in personal construct psychology, vol 1. JAI, Greenwich, pp 203–230
Asimov D (1985) The grand tour: a tool for viewing multidimensional data. SIAM J Sci Stat Comput 6:128–143
Baker AR (1976) Changes in the later middle ages. In: Darby H (ed) A new historical geography of England before 1600. Cambridge University Press, Cambridge, pp 186–247
Bodenhamer D (2010) The potential of spatial humanities. In: Bodenhamer DJ, Corrigan J, Harris TM (eds) The spatial humanities: GIS and the future of humanities scholarship. Indiana University Press, Bloomington, pp 14–30
Bruno G (2002) Atlas of emotion: journeys in art, architecture, and film. Verso, New York
Buja A, Cook D, Swayne D (1996) Interactive high-dimensional data visualization. J Comput Graph Stat 5(1):78–99
Caquard S (2008) Cartographic cinema. Cartographica 43(1):73–75
Cartwright W, Miles A, Morris B, Vaughan L, Yuille J (2008) Developing concepts for an affective atlas. In: Moore A, Drecki I (eds) Geospatial vision: new dimensions in cartography; selected papers from the 4th national cartographic conference GeoCart2008, New Zealand. Lecture notes in geoinformation and cartography. Springer, Berlin/Heidelberg, pp 219–234
Cassirer E (1923–1929) Philosophy of symbolic forms (English translation 1953–1957), vol 3. Yale University Press, New Haven
Chen J (2005) The physical foundation Of economics: an analytical thermodynamic theory. World Scientific, London, pp 16–31. Chapter The entropy theory of value
Chernoff H (1973) The use of faces to represent points in k-dimensional space graphically. J Am Stat Assoc 68(342):361–368
Cilliers P (1998) Complexity and postmodernism: understanding complex systems. Routledge, London/New York
Cleveland WS, McGill ME (1988) Dynamic graphics for statistics. Wadsworth and Brookes/Cole, Belmont
Cohen P (2010) Digitally mapping the republic of letters. ArtsBeat Blog. http://artsbeat.blogs.nytimes.com/2010/11/16/digitally-mapping-the-republic-of-letters/. Accessed 25 Apr 2012
Cook D, Buja A, Cabrera J, Hurley C (1995) Grand tour and projection pursuit. Comput. Graph. Stat. 4(3):155–172
Davis R, Shrobe H, Szolovits P (1993) What is a knowledge representation? AI Mag 14(1):17–33
Dorling D (1994) Cartograms for human geography. In: Hearnshaw HM, Unwin DJ (eds) Visualization in geographical information systems. Wiley, Chichester, pp 85–102
Dykes JA (1997) Exploring spatial data representation with dynamic graphics. Comput Geosci 23(4):345–370
Frank A (1998) Different types of times in GIS. In: Egenhofer MJ, Golledge RG (eds) Spatial and temporal reasoning in GIS. Oxford University Press, New York, pp 40–62
Fyfe D, Holdsworth DW, Weaver C (2009) Historical GIS and visualization: insights from three hotel guest registers in central pennsylvania, 1888–1897. Soc Sci Comput Rev 27(3):348–362
Golledge R, Stimson R (1987) Analytical behavioural geography. Croom Helm, London, pp 52–83. Chapter Spatial Cognition
Goodchild MF (2008) Combining space and time: new potential for temporal GIS. In: Knowles AK (ed) Placing history. How maps, spatial data, and GIS are changing historical scholarship. ESRI, Redlands, pp 179–197
Goodchild M, Janelle D (2010) Toward critical spatial thinking in the social sciences and humanities. GeoJournal 75(1):3–13
Goodchild M, Anselin L, Deichmann U (1993) A framework for the areal interpolation of socioeconomic data. Environ Plan A 25(3):383–397
Gregory IS, Ell PS (2007) Historical GIS: technologies, methodologies and scholarship. Cambridge Studies in Historical Geography/Cambridge University Press, Cambridge
Gruber T (1992) What is ontology? http://www-ksl.stanford.edu/kst/what-is-an-ontology.html. Accessed 23 Apr 2012
Han J (2002) Characteristic rules. In: Kloegen W, Zytkow J (eds) Handbook of data mining and knowledge discovery. Oxford University Press, Oxford, pp 339–344

Harley J (1989) Deconstructing the map. Cartographica 26(2):1–20
Harvey D (2003[1969]) Explanation in geography. Rawat, New Delhi. Reprint, originally published by Edward Arnold and St Martin's Press, 1969
Haslett J, Bradley R, Craig P, Unwin A, Wills G (1991) Dynamic graphics for exploring spatial data with application to locating global and local anomalies. Am Stat 45(3):234–242
Hornsby K, Egenhofer M, Hayes P (1999) Modeling cyclic change. In: Chen P, Embley D, Kouloumdjian J, Liddle S, Roddick J (eds) Advances in conceptual modeling. Lecture notes in computer science, vol 1227. Springer, Berlin/Heidelberg, pp 98–109
Huffaker B, Nemeth E, Claffy K (1999) Tools to visualize the internet multicast backbone. In: Inet'99 Proceedings, San Jose. http://www.isoc.org/inet99/proceedings/4e/4e_3.htm
Hunter GJ, Wachowicz M, Bregt A (2002) Understanding spatial data usability. Data Sci J 2:79–89
Inselberg A (1985) The plane with parallel coordinates. Vis Comput 1:69–97
Inselberg A, Avidan T (1999) The automated multidimensional detective. In: Proceedings 1999 IEEE symposium on information visualization (InfoVis'99), IEEE, San Francisco, pp 112–119
Iturrioz T, Wachowicz M (2010) An artistic perspective for affective cartography. In: Kriz K, Cartwright W, Hurni L, Cartwright W, Gartner G, Meng L, Peterson MP (eds) Mapping different geographies. Lecture notes in geoinformation and cartography. Springer, Berlin/Heidelberg, pp 79–92
Iturrioz T, Cano J, Wachowicz M (2009) Mapping letters through interactive design. Cartogr J 46(4):350–359
Jerding DF, Stasko JT (1998) The information mural: a technique for displaying and navigating large information spaces. IEEE Trans Vis Comput Graph 4(3):257–271
Kaneiwal K, Iwazume M, Fukuda K (2007) An upper ontology for event classifications and relations. In: Proceedings of the twentieth australian joint conference on artificial intelligence. Lecture notes in computer science, vol. 4830. Springer, Berlin/Heidelberg, pp 394–403
Keim D, Kriegel HP (1994) VisDB: database exploration using multidimensional visualization. Comput Graph Appl 14(5):44–49
Kelly GA (1955) The psychology of personal constructs, vol 2. W.W. Norton, New York
Knowles AK (2005) Emerging trends in historical GIS. Hist. Geogr 33:7–13
Kraak MJ, Ormeling F (2003) Cartography: visualization of geospatial data, 2nd edn. Prentice Hall, Harlow
Lackoff G (1987) Woman, fire, and dangerous things: what categories reveal about the mind. University of Chicago Press, Chicago
Langran G (1993) Time in geographic information systems. Taylor and Francis, London
MacEachren AM (1995) How maps work: representation, visualization and design. Guilford Press, New York/London
Masyn K, Henderson C, Greenbaum P (2010) Exploring the latent structures of psychological constructs in social development using the dimensional-categorical spectrum. Soc Dev 19(3):470–493
McIntosh J, De Lozier G, Cantrell J, Yuan M (2011) Towards a narrative GIS. In: Digital humanities 2011. http://dh2011abstracts.stanford.edu/xtf/view?docId=tei/ab-388.xml;query=;brand=default. Accessed 27 Oct 2011
Mitchell TM (1997) Machine learning. McGraw Hill, New York
Newell A (1982) The knowledge level. Artif Intell 18(1):87–127
Owens JB (2007) Toward a geographically-integrated, connected world history: employing geographic information systems (GIS). Hist Compass 5(6):2014–2040
Owens JB (2010) Understanding the impact of nonlinear dynamics on the processes of human systems. White paper published online as part of the National Science Foundation's "SBE 2020: future research in the social, Behavioral and Economic Sciences". http://www.nsf.gov/sbe/sbe_2020/index.cfm. Accessed 2 June 2011
Oyama S (2000) Evolution's eye: a systems view of the biology-culture divide. Duke University Press, Durham/London
Peterson MP (1999) Active legends for interactive cartographic animation. Int J Geogr Inf Sci 13(4):375–384

Piatti B, Bär HR, Reuschel AK, Hurni L, Cartwright W (2009) Mapping literature: towards a geography of fiction. In: Cartwright W, Gartner G, Meng L, Peterson MP (eds) Cartography and art. Lecture notes in geoinformation and Cartography, Springer, Berlin/Heidelberg, pp 177–192

Pickett RM, Grinstein GG (1988) Iconographic displays for visualizing multidimensional data. In: Proceedings of the IEEE conference on systems, man and cybernetics, Beijing/Shenyang, pp 514–519

Polónia A, Barros A, Nogueira M (2010) "Now and then, here and there ... on business": mapping social/trade networks on first global age. In: Kriz K, Cartwright W, Hurni L, Cartwright W, Gartner G, Meng L, Peterson MP (eds) Mapping different geographies. Lecture notes in geoinformation and cartography. Springer, Berlin/Heidelberg, pp 105–128

Robertson GG (1991) Cone trees: animated 3D visualization of hierarchical information. In: Proceedings of the ACM CHI'91. ACM, New Orleans, pp 189–194

Rumelhart D, Norman D (1985) Representation of knowledge: a reader. In: Aitkenhead AM, Slack JM (eds) Issues in cognitive modeling. Erlbaum, London, pp 15–62

Skupin A, Fabrikant S (2003) Spatialization methods: a cartographic research agenda for non-geographic information visualization. Cartogr Geogr Inf Sci 30(2):99–119

Southall H (2003) Defining and identifying the roles of geographic references within text: examples from the Great Britain historical GIS project. In: Proceedings of the HLTNAACL 2003 workshop on analysis of geographic references. Edmonton, Canada, pp 69–78

Southall H, von Lunen A, Aucott P (2009) On the organisation of geographical knowledge: data models for gazetteers and historical GIS. In: 5th IEEE international conference on E-science workshops, Oxford, pp 162–166. doi: 10.1109/ESCIW.2009.5407970

Villuga J (1546) Repeortorio de todos los caminos de España. Pedro de Castro for Juan de Espinosa, Medina del Campo

Warf B, Arias S (2009) The spatial turn: interdisciplinary perspectives. Routledge, London

Wooldridge M (2000) Reasoning about rational agents. MIT, Cambridge, MA

Zhang R (2005) Hierarchies for event-based modeling of geographic phenomena. Electronic theses and dissertations, paper 568. http://digitalcommons.library.umaine.edu/etd/568. Accessed 27 Oct 2011

Chapter 10
Writing Visual Histories: An Interview with David J. Staley

Charles Travis and David J. Staley

David J. Staley is currently the director of The Harvey Goldberg Center for Excellence in Teaching, in the Department of History at The Ohio State University. He is also the past executive director of American Association for History and Computing (2003–2008). He possesses a doctorate in German history, as well as a master's and bachelor's degree in history, with a minor in economics *cum laude* from Ohio State University, and has recently published a work entitled *History and Future: Using Historical Thinking to Imagine the Future* (2007). In his career as a historian, Staley has engaged various types of visualization and digital technologies. As the spring tides of the twenty-first century's "digital revolution" were beginning to flow into the corridors of academia, Staley observed: "Art historians of the future may very well define ours as one of the most important in the entire history of images, on par with Impressionism, Cubism or the Baroque" (Staley 2002). He noted "the images of our period are composed not with marble or oil on canvas but with pixels and bits of electronic information" (Staley 2002). Instead of the grand motifs and subjects conveyed by classical and modern art, new media imparted images concerned with "relatively mundane topics such as the structure of social relations, the rhythms of the heart or the landscape of numbers" (Staley 2002).

As author of the groundbreaking text, *Computers, Visualization, and History: How New Technology Will Transform Our Understanding of the Past* (2003), Staley observed that "geographic information systems, data-rich schematics, three-dimensional information spaces and virtual reality graphics have become commonplace" in the "symbolic landscape" (Staley 2003a, 3) of historical writing, research, and study. Considering nineteenth-century British historian Thomas Carlyle's aphorism that "narrative is linear, action is solid" (Staley 2003b, 40), Staley has written that "many of the events of the past [...] might be better understood

C. Travis (✉)
Trinity Long Room Hub, Trinity College, University of Dublin, Dublin, Ireland
e-mail: ctravis@tcd.ie

D.J. Staley
Department of History, Ohio State University, Columbus, Ohio, USA

A. von Lünen and C. Travis (eds.), *History and GIS: Epistemologies, Considerations and Reflections*, DOI 10.1007/978-94-007-5009-8_10,

as a three-dimensional structure extending in breadth and depth as well as length" (Staley 2003b, 41). Staley argues that complex "historical phenomena does not occur one after another like a line of words but more like the acoustic space of symphony, with events occurring simultaneously, located in a variety of places, moving at different rhythms of time" (Staley 2003b, 41). In a piece entitled *Finding narratives of time and space* (2007), Staley further addresses the potential for GIS's multidimensional engagements with history. This technology, Staley contends, does not simply produce "attractive visual aids to illustrate a textual presentation" but "tools to explore data" (Staley 2007, 37). He notes that one result of the "digital revolution" is that "text on the computer screen has challenged many of our ideas about linear narrative" (Staley 2007, 42). Furthermore, "hypertext narratives are often organized like maps; these stories exhibit spatial qualities, as they unfold in nonlinear or multilinear ways" (Staley 2007, 42) and that "GIS like hypertext, alters the relationship between map creator and map viewer" (Staley 2007, 45). Staley argues that a GIS map appears as a rich expression of "ergodic literature" (a work that requires labor from the reader—or viewer—to create a path; Staley 2007, 45). Employing GIS to "write" and "narrate" cartographic histories "lets the mapmaker depicts change and movement through time and tell a story chronologically" (Staley 2007, 39). This provides a "space" to operationalize Carlyle's notions about the breadth, depth, and length of history. Staley observes, "the procedural author of a GIS space serves as a kind of 'narrative architect'" (Staley 2007, 45). He concludes, "far from the classic Aristotelian linear narrative, a GIS is a multidimensional narrative space of images and words . . . of story and interactivity" (Staley 2007, 45).

In the spirit of the "digital humanities," the following interview with David J. Staley was conducted electronically.

10.1 On Geography

CT: How do you regard geography's role in history? Does it inform history; is it a discursive framework or just a nominal scope (i.e., something to define the geographical scope of a historical study, but not with an actual value to the study)?

DJS: I think, generally speaking, this latter description is probably closest to the mark. If by "value" you mean a central methodological feature of our work, then I don't think we can make this claim for all historians. There are, of course, subspecialties for which geography looms large. The Annales school wanted to unite geography and history, in order to be mindful of the role of the "stage" upon which history is enacted. Military historians are interested in space and terrain, as are diplomatic historians. I think historians tacitly assume that knowledge of geography spatially frames our work; it certainly informs our teaching, where the "map quiz" is a regular feature of our classrooms. But as a methodological category, I don't know how widely practiced geographical methods are in our work?

CT: Do you follow the *spatial turn* discussion in history?

DJS: A little, although I must admit to being skeptical about this phrase being applied to history just yet. A "turn" to me implies a paradigmatic shift in methodological orientation. Historians are notoriously methodologically conservative, and so any talk of a "turn" usually unfolds over a long period. I think there was, for example, a "gender turn" in history such that gender as a category of historical investigation has become paradigmatic; at this year's AHA convention, 192 presentations list "gender" as the defining category. By contrast, 26 presentations identify "spatial" or "GIS" as a category (although I am impressed by this number!). I know there are a growing number of historians doing very good work uniting geography and history (under the influence of GIS and other digital methods). However, I think we are, at best, in only the very earliest days of a "turn" toward space as a paradigmatic category of historical investigation. There were those historians, in the 1970s and 1980s, who championed quantitative methods. And indeed, there are today some historians who rely on statistics to conduct their work, usually clustered around journals like *Social Science History*. I suspect that, had the phrase been available to them, these historians would have announced a "quantitative turn" in history. But quantitative methods never really caught on with historians at large, such that these methods became a regular part of every graduate student's training in history. Gender became paradigmatic; quantitative methods did not. Whether geography becomes one of these paradigmatic categories has yet to be seen, which is why I am hesitate to announce such a turn.

10.2 On Computers and GIS in History

CT: Specifically, could you describe and illustrate how GIS or computer applications in general have provided new insights into your own historical work?

DJS: My interest in computing and history has long focused on the implications of these tools for how we represent knowledge of the past. At around the same time, most historians were learning about computers as word processors, around the mid-1980s, I was being introduced to the graphics capabilities of these tools. I was taking courses at that time in the nonlinear sciences, chaos theory and complexity science, and was struck by how computer graphics were central to their methods, not as pretty pictures but as a way to uncover and represent patterns heretofore unseen by scientists and mathematicians. I also received my first visit to Ohio State's Advanced Computing Center for the Arts and Design (ACCAD) and saw firsthand how the new generation of computer graphics was being used by both computer scientists and artists. Seeing how computer graphics were being used in these disciplines, I announced (to no one in particular) that the visualization capacity of computers should be applied to historical representation—which became the theoretical basis of my first book, *Computers, Visualization and History*. From this grew my interest in spatial patterns, whether those were geo-referenced or, more interestingly to me, abstract or nonrepresentational patterns. Clearly, when I was introduced to them, GIS tools were part of this concern for representation; I argued that GIS displays offer interesting possibilities for how we might represent the past, beyond what can be written in an article or monograph, and not merely as an illustration to arguments made in words. In the book, I envisioned a day when a graduate student's dissertation in history could be composed of an elaborate and rich GIS map, with a short written "designer's statement" to illustrate the argument made via spatial patterns. We are still at some distance from this vision, I fear. But I remain interested in using computers for visual representations.

CT: In *Computers, Visualization, and History* (2003), you write "like prose, visualization is a template for ideas, a means of ordering one's thoughts about a complex subject." GIS, as it is generally conceived and applied, frames Euclidian and Cartesian senses of space. Do you think that such a structuring of space is more amenable to certain schools and historical approaches? In this regard, I am wondering about the types of historical narratives that such a structuring of space could hide, or perhaps elide?

DJS: Yes, I think so. I have been fascinated by work my colleagues have been undertaking with 3-D modeling, of places such as ancient Rome and Hadrian's Villa, the Chicago Exposition of 1892, and the spaces of Dublin during the War of Independence. What I find striking about this work is not only the elaborate architectural reconstructions (which are indeed become more and more realistic) but the larger historiographic theme of the recreation and "experience" of historical space. What unites much of this work is the idea that reconstructing these spaces and making them virtually habitable and maneuverable for an avatar allows us to recover and represent spaces from the past. In this sense, space itself becomes an historical category; it is historical in the sense that space is not uniform, and is the product of unique local circumstances and historical trajectories. I am in the very early stages right now of a project with a colleague who is studying department stores along a boulevard in early twentieth-century Buenos Aires and is especially interested in the gendering of such spaces. My first impulse upon reading a proposal of her work was to say "we need to virtually model this space, and confront an avatar moving through the space with that 'feeling of gendered space' that you are arguing." But quite apart from these kinds of reconstructions of physical, Euclidean spaces, I am even more interested in various kinds of conceptual spaces that we as historians might also explore. I am especially interested in what I like to term "Playfairan space." By this, I mean William Playfair, whose charts and graphs Edward Tufte has identified as some of the first modern data graphics. Playfair's graphs "mapped" an abstract space whose axes were bounded by qualities/quantities as opposed to physical dimensions. So, for example, in charting imports and exports in England over time, Playfair was creating a "space" of data and information. Mapping, of course, is not and should not be confined to physical, Euclidean spaces. What I find striking about the work of Franco Moretti, for example, is not only his mapping of the physical spaces of a novel but his mappings of the rise and fall of genres of the British novel over time.[1] Moretti has identified (or perhaps invented?) and explored a conceptual space. As an historian, I am becoming ever more interested in developing such conceptual spaces and mapping their properties; I am planning right now a new work that will take the entire corpus of the *American Historical Review*, text mine it, then map out qualities such as word frequencies in order to develop a map of shifting historiographic interests among historians (the word/concept "gender" over time, or "space" over time). In a sense, I would be seeking to map different "turns" in American historical thought and wish to array these in a conceptual space. I am working right now on a text visualization called "writing space III" that arrays blocks of text in a three-dimensional space. "Reading" in this environment means not just moving the eyes across the page but also moving the body (the metaphorical body, represented by the first-person avatar) through a 3-D conceptual space. In this manner, spatial history, it seems to me, can mean much more than what we usually think of as "historical geography."

CT: Furthermore, in *Computers, Visualization, and History* (2003), you make a case for "visualization" as a historical visualization tool to counter the discipline's preference for Aristotelian, linear, and narrative presentations. Can you expand on this, perhaps by considering how various forms of "visualization" might produce alternative historical methodologies?

DJS: If historians claim that their representations "map" some reality of the past, I simply ask historians to more closely align their representations to the "thing" being represented. Often, what we represent is a complex, multidimensional reality. I would like to see historians rely on representations that are similarly multidimensional. Historians are very logocentric, even when faced with a problem that cries out for another form of representation. I use this simple example to demonstrate what I mean. Take a standard road map of the United States, then "translate" all of the information displayed on that map in written form: "New York is northeast of Philadelphia which is north of Baltimore, all of these cities on the coast of the Atlantic Ocean, connected by US Route 1..." Clearly, a map is a representation that makes more sense here than a written account. That is because the information being represented here is simultaneous. I suspect there are many such instances in history where some other form of representation other than a written

[1] Cf. Moretti (2007).

prose account makes more sense, yet historians still privilege the written account as "serious history." This is what interests me about spatial history and GIS generally: that mapping aspects of the past more closely aligns the representation to that being represented.

CT: Both proprietary and open-source geospatial technologies (as well as other computer/digital softwares and hardwares) are obviously embedded in and reflective of political/economic/social/ideological structures and networks. How do you think academic researchers who engage in such technologies should situate and contextualize their work in relation to these structures and networks?

DJS: Wow, very knotty problem, isn't it! I know that one debate I have with colleagues is whether we should publish in any journal controlled by one of the large for-profit companies like Elsevier. (I recently dealt with this issue myself). The journal is one of the tops in the field, and one must weigh the prestige of publishing under those conditions versus publishing under the newly emerging spirit of open-source publishing (as someone who works at a noted land-grant university, our first impulse is to make our scholarship widely and freely available.) The issue also extends, of course, to the kinds of software we are using in our work. When available, I always encourage my students to use open-source applications, but with the understanding that the needs of the problem, the research agenda, have primacy.

CT: In *Computers, Visualization, and History*, you briefly discuss GIS as a means to explore counterfactual histories and alternative scenarios.[2] Niall Ferguson has made some hay in this field of speculation with his edited volume *Virtual History*.[3] One of the elisions in Ferguson's volume seems to be the lack of any substantive discussion concerning role of geography in shaping counterfactual histories. Though you mention counterfactuals briefly in your 2003 volume, could you expand on your thoughts concerning the role of GIS in exploring counterfactual histories and alternative scenarios.

DJS: I once used as an examination question in an American history class two maps from Donald Meinig's *The Shaping of America*. These were two "counterfactual" renderings of nineteenth-century American history, one a "greater United States" which had expanded into Mexico and around the islands of the Gulf of Mexico including Cuba, in effect turning it into an American lake. The second was a "lesser United States," with the confederacy split-off from the United States, Texas, California, and the Mormon state of Deseret as separate countries. The examination question I asked my students was "what would have had to happen in each case to produce the results in these two maps?" The students would have answered something about the decision to claim larger amounts of territory as part of the settlement at the end of the Mexican-American War or would have

[2] Staley (2003a, 125).

[3] Cf. Ferguson (1997).

written about an honorable settlement of the Civil War in a scenario where the Confederacy receive material support from Britain and France. That is, to understand the narrative of each map, students would have had to have deep knowledge of the possibilities that existed in nineteenth-century American history and be able to express something of the contingent nature of historical causation. In order for students to really understand these possibilities and to understand the nuances of causation, I have found that mappings like these can be a powerful means to visualize historical counterfactuals. Some researchers use GIS in order to explore alternative future scenarios, along the lines of "if we alter this variable, how would the system change?" or "if we built a highway here, what would be the environmental impact?" It strikes me that historians might also explore counterfactuals by doing something similar with our historical data: alter a variable, then discover how the system so visualized on the map would change.

CT: Your piece "Finding Narratives of Time and Space" in *Understanding Place: GIS and Mapping Across the Curriculum* (2007) draws upon critical studies, in addition to literary and cyber-text theories to consider possible GIS approaches within the humanities. Can you discuss how these disciplines, and any others, may inform and shape your practice of historical investigation and teaching?

DJS: While I certainly draw upon these sources to situate my reading of GIS in the humanities, I suppose I am more pragmatic than theoretical in my practice.

CT: Finally, should computer scientists—with the assistance of historians—develop better (i.e. more accessible) software? Or should historians become *bricoleurs* and more technically proficient to make better use of the tools/software available to them?

DJS: The simple answer is yes! I am thinking here specifically of ArcGIS, which still (as I understand it) does not have a useful animation feature in order to show change through time, one of the hallmarks of historical thinking. If programmers and historians could work together to produce something like that, that would be a boon for historians and for the spatial humanities generally.[4] As for whether historians need to become more technically proficient, I think the answer is yes, although I would limit this demand to those engaged in digital history, not historians as a whole. That is, if one is going to work in digital history (whether with GIS or not), technical proficiency is as necessary as a working knowledge of Latin and German are to a medieval historian.

[4]It should be noted that ESRI has released ArcGIS10 in the summer of 2010, which introduced a feature to animate temporal data, among other things. This is not yet widely known in the historical GIS community.

References

Ferguson N (ed) (1997) Virtual history: alternatives and counterfactuals. Picador, London
Moretti F (2007) Graphs, maps, trees: abstract models for literary history. Verso, London
Staley D (2002) Visualization-ism: an art history. Interface 2(6). http://bcis.pacificu.edu/journal/article.php?id=491. Accessed 28 Apr 2012
Staley D (2003a) Computers, visualization, and history: how new technology will transform our understanding of the past. Sharpe, London
Staley D (2003b) Computers, visualization, and the representation of history. Hist Speak 4(4):40–42
Staley D (2007) Finding narratives of time and space. In: Sinton D, Lund J (eds) Understanding place: GIS and mapping across the curriculum. ESRI, Redlands, pp 35–48

Chapter 11
GIS and Research Into Historical "Spaces of Practice": Overcoming the Epistemological Barriers

Sam Griffiths

11.1 Historians and HGIS

"Mainstream" historians have not been pioneers in developing HGIS applications. This should not be a cause for surprise. Gregory and Ell—historical geographers and authorities in the field—have commented on the difficulties experienced by non-specialist users of GIS in overcoming disciplinary boundaries even within the subdisciplines of geography itself (Gregory and Ell 2007, 19ff). They explain how the relationship of GIS to geography and HGIS to historical geography—its most natural disciplinary "home"—is still being negotiated. The inevitable emphasis on technical proficiency, spatial analysis and the implicit theorization of space as a smooth Euclidean surface causes no fewer (possible more) problems for many human geographers as it does for historians, many of whom have embraced a more qualitative approach to space as "place" (Baker 2003, 219). Gregory and Ell go on to (rightly) assert that the "historian wanting to use GIS must not only learn the technical skills of GIS, but also must learn the academic skills of a geographer" (Gregory and Ell 2007, 11). This is a lot to ask—a point developed by Travis in his chapter in this volume. It suggests that those specialists with an interest in how historians use GIS should not assume that it is only a matter of recalcitrance or ignorance on behalf of historians that there is not a wider uptake. Quite apart from the question of relevance of HGIS to the research itself, there may well be good disciplinary reasons for exercising caution in deploying this technology that run deeper than simply a lack of familiarity with some of the key concepts and expertise with the technology.

Having said that, there *are* undoubtedly good reasons why some historians would do well to engage more with HGIS, not least because it opens a methodological window onto an interesting but difficult to research historical problem—the relationship

S. Griffiths (✉)
Faculty of the Built Environment, University College London, London, UK
e-mail: sam.griffiths@ucl.ac.uk

A. von Lünen and C. Travis (eds.), *History and GIS: Epistemologies, Considerations and Reflections*, DOI 10.1007/978-94-007-5009-8_11,

of society and space. Urban historians, for example, have long acknowledged the difficulties in tracing "patterns on the ground to patterns in society" (Cannadine 1977, 1982). The challenge of relating the morphological structure of a settlement to the lives and culture of the people who live there is a particularly pressing problem (Pooley 2000). This issue cannot simply be resolved by training historians in the use of mapping software but requires an exercise in theory building and reflection on the epistemology of HGIS itself as a dedicated method intended to help historians to see and think spatially in new ways. In practice however, the emphasis of historians engaging with the "spatial turn" has been less on the historical geography itself than on the nature and meaning of spatial representation. There is something of an irony here since most academic history prioritizes archive-based empirical work above theory, but when it comes to the question of "mapped data", the prevailing focus is almost entirely on the normative context of their "production" as cultural objects rather than on their substantive content. While not disputing the importance of this emphasis—the proposition here is certainly *not* that maps are "true" representations of reality—this chapter advances an alternative position that a too-exclusive focus on historical space as representation has created something of an epistemological blind spot for historians wishing to access and substantively describe "spaces of practice" produced by everyday activity. However, any difficulties for historians with the "objectivity" claimed for spatial analysis in HGIS are secondary to the greater issue that in historical research it is the source materials themselves that generally constitute the epistemological framework for historical explanation. More than anything, the application of a particular method has to be meaningful in terms of that framework, it must resist at all costs charges of "manufacturing" or manipulating the evidence. While few contemporary historians would subscribe to the argument that their work simply "reveals" the past "as it actually was", much historical research exists in a philosophical "grey" zone in which the sources and the historian's own contextual knowledge of the period act as powerful constraints on the construction of the historical account (Gardiner 1961). Claims for "objectivity" lie in the coherence of the historian's account in terms of the sources and other accounts, rather than by appeal to methodology as such. From this perspective, well-defined specialist methods such as GIS risk introducing a distorting intermediary between the historian and his or her source materials. However, as other contributions to this volume establish, there are alternative ways of imagining the mediation of an HGIS that might be more productive for historians.

History does not regard itself as a methodologically driven discipline in the sense of allowing the capability of a particular method to shape the questions asked, the sources consulted and the conclusions reached. This is not, of course, to assert that other disciplines seek to do this, but the concern of the historian with the human past gives this concern a definite resonance. On the whole, historians are concerned with the specificity of particular times and particular places. They are accustomed to bring as many extant sources as they can to advance their account—privileging none and certainly not manufacturing additional ones (which would be fraudulent in this context). Well-defined scientific methodologies such as GIS, on the other hand, tend to have roots in theories of methodology (or epistemology) that aspire to generalization—that is, to arrive at "models" of general principles and laws. From this perspective, there is no a priori problem with a method intervening to artificially "produce" the phenomena that are the test of the theory.[1] In historical studies, by contrast, the sources come first and last, with "scientific" generalizations constrained within the narrative and therefore left contingent and open (Gardiner 1961). The need for methodological expertise beyond the historian's own knowledge of the period (e.g. linguistic or statistical skills) is likely to arise organically through the need to read and interpret particular sources. In general, the sources come before the specialist "method" which has possibly been mastered only so far as is required for dealing with the source material in question.

We might summarize by saying that historians specialize in understanding sources—that is, traces and remnants of the past by a process of placing them in appropriate historical contexts—rather than in methodological expertise per se. The strength of this approach is that the specificity of the research is not made secondary to the requirements of the methods employed and that a plurality of both source materials and interpretative approaches can be used. While this position may be difficult to defend philosophically—because historians cannot ultimately deny that they are employing a method of one kind or another—it nonetheless serves the study of history by putting the focus on the sources. The weakness is that such an ad hoc approach can licence a certain methodological naivety and subsequent on lack of reflection on the historical method by historians themselves (cf. Gunn and Faire 2012), perhaps why so many theoreticians of history are not actually practising historians. It also promotes a certain conservatism of method—not necessarily a bad thing in itself—but one which can be limiting when an extra level of methodological intervention in the sources is required to get them to "speak". When the sources in question are geographical or architectural ones (e.g. maps, plans and surveys) as might typically be expected of research conducted in an HGIS, then historians could usefully contribute to thinking about what kind of intervention in the source material this method actually constitutes in terms of their own practice. In presenting itself as a useful "tool"—as proponents of HGIS often do—they avoid critical engagement with its epistemology. Working across disciplines, researchers are always likely to be highly sceptical of a method that presents itself as a "black box", effectively a button to press, that once they have learned some basic technical skills will enable them to produce the desired outputs. First and foremost, historians needs to understand whether and in what sense HGIS can advance his or her understanding of the sources. This is not achieved by applying a well-honed method or "procedure" to appropriately customized data in a generic fashion—a move which threatens to take the historian out of the process and produce decontextualized "results".

[1] This interesting process that the philosopher of science Hacking (1983) calls "the creation of phenomena" will be returned to later in this chapter.

11.2 HGIS and the "Spatial Turn" in Historical Research

The widely discussed "spatial turn" in social theory has made a strong impact in many areas of historical studies.[2] Tracing their inspiration back to Lefebvre's groundbreaking theoretical work but also Bourdieu, Giddens and Harvey, the idea that space is not a passive "container" of social life but is itself fundamentally constitutive of social life and a therefore legitimate object of critical enquiry has enabled historians to open a new frontier of research in which geographical notions of "space", "place" and "location" are foregrounded.[3] Typically, this research strongly resists the notion that built environment (broadly interpreted as the "human environment") and particularly its representations in art, cartography and literature can in any sense be regarded as "natural" or "neutral". Of particular interest therefore are the history of hegemonic systems of governance and representational systems associated with the symbolic "ordering" of space (e.g. Joyce 2003).

Given this theoretical trajectory, it is perhaps inevitable that the application of GIS might seem unattractive to many historians—even putting logistical and technical issues to one side. It is not simply a question that—as Gregory and Ell (2007, 19) acknowledge—HGIS privileges the representation of patterns and relationships in space without identifying the "underlying processes" that produced them (though we should not forget that identifying patterns and relationships is not unimportant). It is also that GIS is itself one of the repertoire of techniques for the production of what Lefebvre calls "abstract space" that is at once hegemonic, systematizing and annihilating of difference. The systematized cartographic "ordering" and "representation" that GIS enables cannot simply be regarded instrumentally when its genesis is clearly associated with so many political agendas. In general, this pertains to strategies for state surveillance and regulation of territory. The large number of GIS projects concerned with the online publication and promotion of "national data" (historical or otherwise) cannot entirely escape this ideological association with the production of national or state "space", however pragmatically useful they may be. This is treacherous ground for non-specialists whose introduction to the "spatial turn" came via the Lefebvrian tradition. Wherever a method appears to impose something—an order or selective preference—on the diversity of historical materials, historians are likely to be suspicious.

Unfortunately, the HGIS community does not receive much epistemological support from its disciplinary partners in human geography, symptomatic of a perhaps surprising lack of dialogue between these fields. Certainly, human geographers such as Massey (2005) are among the most critical of methods that appear to prioritize (or "fetishize") the projection of static images of the social onto continuous Euclidean space at the cost of examining the processes that lead to such (hegemonic) images being produced. The particular bugbears here are the

---

[2]Cf. Kingston (2010), Biernacki and Jordan (2002) and Gunn and Morris (2001).

[3]Cf. Lefebvre (1991 [1974]), Bourdieu (1990), Giddens (1984) and Harvey (1973).

synchronous nature of geographical representations that, she argues, have the effect of depriving space of its multiplicity and temporal dynamism. Thrift (2008), among others, has taken the geographical theory of space into the distinctly non-Euclidean domain of imbricated and fragmented networks where the authority of any given mapping is always suspect. Other lively areas of research, including the "mobilities paradigm", and an approach to material cultures informed by phenomenology tend to be highly critical of traditional quantitative or cognitive approaches to space, often (though not exclusively) preferring a more qualitative and experiential emphasis (Urry 2007; Tilley 1994).

Research into historical space, in terms of Lefebvre's well-known tripartite scheme, has tended to concentrate on disclosing abstract "representations of space" or decoding semiotically rich "representational spaces" as "texts". By contrast, "spatial practice" and the "spaces of practice", that is, the histories of the embodied spaces of everyday life, are noticeable by their absence. One reason is that such performative spaces arise in Thrift's "nonrepresentational" domain that may be productive of representations or have representations imposed upon it but is not ultimately reducible to representation. Certainly, there are rich accounts of historical street life,[4] and in Jerram (2011), a fundamental commitment to the idea that the "where" of history is as important as the "when". Nevertheless, while such accounts suggest strong shift to address history from the perspective of spatial practices, there is little overall sense of how "space" itself is theorized in this mode of enquiry. In itself, this need not matter—the quality of these historical accounts speak for themselves. However, one is entitled to ask whether the "spatial approach" to historical research could benefit from such a theorization that would assign rather greater agency to spaces of practice than that of a "setting" in the spirit of Giddens' "locale", where for all his emphasis on space such agency seems to be lacking—a background against which social life is enacted. Such a mode of historical enquiry may have "simple" spatial description as its initial task but with the wider aim of elucidating spaces of practice in relation to the representations of space and representational sign systems that are relatively more intelligible and accessible to archival research.

Historians, no less than human geographers, are concerned that an emphasis on spatial representation, whether as a surface, a network, a grid or configuration, erases historical ("internal") time by eliminating difference. Clearly, a technology such as HGIS, for all its ability to handle longitudinal data and simulate temporal processes, is necessarily limited in how far it can handle temporality in this epistemological sense—the complexity is just too great. The spatialization of time is a serious issue for historians for whom the contingent reconstruction and reading of the complex interplay of historical events are key to their accounts and therefore sits uneasily with the mechanical phases of ("external") Newtonian time—this even before more experiential notions of temporality enter the frame.[5] For example,

[4]For example Shoemaker (2004) and Gunn (2000), or Corfield (1990).

[5]See Griffiths (2008); also Tuan (1978).

although Braudel's scheme of historical time describes three dominant frequencies, it was the geographical *longue-durée* that was clearly his preferred register of explanation (Braudel 1980). Against the powerful "natural" constraints of the environment, the temporalities of everyday life appear weak and episodic. However, this is not to say that "events" were an irrelevance to Braudel as a historian; on the contrary, he devotes a considerable section of his opus *the Mediterranean* to their consideration (Braudel 1976). The point is rather that for Braudel, such events were deprived of any claim to "deep temporality", which is portrayed as the exclusive property of geographical structures that consequently, and perhaps ironically, can appear rather *timeless* by comparison. It is not hard to see how the same synchronic mode of explanation could be translated into HGIS by default, for example, through the identification of discrete topographic, socio-economic and ethnographic "layers". The consequence of "too much" space in this sense is that it is hard to see how historical events themselves cohere and present themselves in time (Bentley 2006). Historical "events" are similarly marginalized by the philosophical tradition that characterizes space either in terms of the subject (mental space) or the object (physical space)—thereby effectively eliminating "social space" that implicates both in a generative (historical) dynamic, as Lefebvre objected. A too-exclusive focus on historical research aimed at disclosing spaces of representation and decoding semiotically privileged representational spaces therefore runs the risk of excluding the embodied field of spatial practice from notions of historical causation that properly belong in the social sphere. This discussion may seem esoteric in the context of HGIS, but it matters because the question of how historical time is inscribed in space is integral to the argument asserting the value of spatial description to historians that is advanced here.

Such a discussion is haunted by the spectre of nineteenth-century environmental determinism, the decaying shells of modernist architectural projects and the effects of the "quantitative revolution" in geography, widely held responsible for denying to space the capacity to be socially meaningful. This means that while other structural categories of the social, one might think of the family, the economy or gender, can be handled sensitively such that it is possible to invoke them without attracting charges of determinism or reductionism—this is not the case with space. The notion of space's "production" in Lefebvre clearly invokes a historical process that involves people both in the representation and *doing* of space (a process fraught with contradictions). Social space is not simply a passive output but also—in some sense—an agent through which social life is brought forth. The difficulty for researchers is that the "common language of the city" referred to by Lefebvre in his evocative description of Venice remains largely opaque (Lefebvre 1991 [1974], 73). This has been noted by the architectural morphologist Hillier who argues that the opacity stems from a reluctance in social theory to assign any agency to the spaces of practice in the belief that this risks repeating the mistakes of the quantitative revolution (Hillier and Vaughan 2007; Hillier 2008). Hillier and Hanson's development of space syntax theory, with its emphasis on the syntactical description of "spatial configuration", is a powerful attempt to address this issue (Hillier and Hanson 1984). Whether one accepts Hillier's approach or not, he is surely right in noting that a relative lack of theoretical interest in "real" or everyday spaces has an impact in how they can be approached in empirical studies (Hillier 2008, 223).

Lefebvre stated that "man does not live by words alone" (Lefebvre 1991 [1974], 35). This evocative phrase suggests how representational forms of meaning cannot be sustained in isolation of their "affordance", to use the psychologist Gibson's term, by the spaces of practice inhabited by the embodied human agent (Gibson 1979). Lefebvre (1991 [1974], 38) articulates such space in terms of its "secretion" through performative spatial practice. Without some notion of spatial description, when the explicitly representational is put to one side, we are still left with the "blank wall", the stage setting, which in itself affords no meaning. Proponents of the spatial turn may object to quantitative description as being reductive, but semiotic systems can be no less abstract in describing spaces of practice. An approach to HGIS more attuned to the requirements of historical research into lives lived in everyday spaces could play a central role in such a project of spatial description by bringing a range of perspectives and sources to bear on identifying modes of temporal inscription in space and their relation to human activity. For this to succeed, HGIS should not be presented as a black box "tool" to be applied but rather as an "open-source" methodology to be critically engaged with and adapted.

11.3 HGIS "Method" or "Tool"?

There are at least pragmatic reasons why a historian might decide to employ an HGIS in his or her research. GIS specialists produce very powerful data visualizations, and good cartographic representations clearly have a place in many historical studies. Most historical source material can, in theory, be linked to a GIS so long as it is associated with a location. The utility of GIS as a "tool" is not at issue here, nor is the question that the considerable technical know-how and disciplinary-specific knowledge required presents a considerable barrier to historians in practice. Such issues, however, can be overcome by hard work and good training. The question that this chapter is concerned with is primarily epistemological. What, if anything, can HGIS offer historians above and beyond the functional utilities mentioned above—what could it have to do with the creation of historical knowledge that would not be available to them otherwise? A reciprocal question should be posed, how could engaging with mainstream historians offer anything to HGIS specialists? In other words, what is the conceptual basis of this disciplinary exchange?

One way in which sociologists, geographers and historians have tried to get themselves off the epistemological hook of the paucity of descriptions of the spaces of practice is to recourse to the often polarized discourses of "space" and "place" (see, e.g. Gieryn 2000, 465). In this context, "space" is routinely conceived as an abstract, continuous surface, quantifiable and ideal, while "place" addresses historical particularity and meaningfulness. Location is another, less loaded, geographical term that refers to a fixed point (or coordinate) in space.

GIS specialists (e.g. in planning) may talk about "placemaking", while phenomenologists will have recourse to the "poetics of space", but the telling fluidity of terminology across disciplines is insufficient to dissolve the conceptual difficulties caused by the dominant theoretical discourses of "space" and "place". Too often, these complacently assume sharply opposed standpoints under the illusion of a "just" division of academic labour. This achieves little more than the hardening of the epistemological boundaries between disciplines exactly where they promise to be most productive, again recalling the distinction between physical space and mental space that Lefebvre sets out to undermine (also Hillier and Leaman 1973). Rather than compartmentalize notions of "space", "place" and "location", historians and geographers should become more sensitive to the sometimes complementary and sometimes incommensurate ways in which all these concepts can be put to work in their research. "Space" and "place" are distinct but not exclusive concepts. Without dictating definitions, notions of "space" are necessary to elucidate those affordances of the built environment that are the domain of perceived space and spatial practice, whereas notions of "place" pertain to those representational forms that are more immediately intelligible in terms of social meaning. Yet, it is hard to conceive the intelligibility of the latter without some notion of the productive, generative capacity of the former. If a technology such as HGIS has any potential for advancing historical knowledge, it lies in providing a platform for opening up a *dialogue* between the images and representations of space that are often textual or highly symbolic in nature and the spaces of everyday life that may only be researchable through maps, plans and artworks but are less accessible to description and interpretation as semiotic systems because, in the first instance, they pertain to the essentially non-representational domain of spatial practice. One has to tread very carefully here, of course, because the kinds of source material available to the historian in this respect, particularly maps and plans of various kinds, are not neutral representations, as has already been acknowledged. However, the historian of the quotidian is not, one thinks, unable to reckon with these limitations in developing their account in dialogue with the GIS.

One might start by considering the map or plan as a characteristic source for analysis in an HGIS. The historian is generally less interested in the map itself than in the social processes it reveals, the map as a representation of hegemony. It was made at a particular time, for a particular purpose and should be understood in that light. Yet, why not approach the map itself critically in terms of what it can tell us about the materialities of everyday life and develop the analytical capacity to assist in generating descriptions of spatial practice that can beg questions and help to provide answers about the role of material life in different historical contexts? Maps are very rich historical sources, not just in what they reveal about the world of the map-maker but also what they can reveal about the lives of the people who inhabited the spaces depicted in the map (and who inhabited the same world). It is curious how it even works at the cutting edge of the history of street-level urban life and material cultures (e.g. Jerram 2011; Guildi 2011) make relatively sparing use of maps and plans—at least in these published works. Geographers and architects are, of course, trained readers of maps and plans, but they do so without the same concern with temporality as the historian. HGIS has a role to play in developing an epistemology of space-place description in which different spatialities are not approached in isolation but as mutually constitutive and temporally dynamic.

11.4 HGIS as a Mediator of Historical Source Materials

Any well-defined specialized method such as GIS cannot be said to mediate innocently between the historian and the subject of his or her study. It is therefore important to know what kind of method or tool it is since these questions often go unasked within disciplinary boundaries. On the whole, it has been argued that historians will allow the sources themselves to guide them to appropriate methods. This approach engenders an understandable suspicion of a "tool set" that takes over the "black box" mediating in the critical dialogue between the historian and the sources. Writing in an introductory survey of historical research methods, the historical geographer Lilley notes that "as with any technique or method, GIS is a means to an end and not an end in itself in historical study" (Lilley 2012, 136). In this comment, Lilley shows sensitivity to the dislike of many historians for methodologically driven approaches that can seem to subsume the research process.

There are three aspects of the "black box" that cause historians particular concern. The first of these relates to *reductionism*, adjusting diverse historical source material to fit a predetermined schema or metadata model that seeks to isolate variables rather than doing justice to the messy and fragmentary nature of the sources. Related to this is the need to organize data into a restricted range of categories, standardize place names and other data, stretch maps to equivalent projections and create arbitrary synchronic layers by bracketing sources within a particular temporal range into broad chronological bands for the purposes of representation. Such systematic methodological transformation of historical sources completely changes the nature of what it is possible to do—both destroying some forms of historical knowledge while, possibly, giving rise to others—a process that needs to be better understood. Although one could fairly argue that any process of historical research entails such a process of selection, it is the sense that the history is serving the method, rather than the other way around, that historians would naturally seek to avoid.

The second pertains to the suspect *authenticity* of the representation as the final "output" of an HGIS model. Historians are used to returning to their own sources again and again in writing their historical account. The disciplinary tendency of GIS specialists to regard their visualizations as the definitive "end point" of a linear analytical process can seem contrary to this discursive and open-ended approach. Two further difficulties with GIS visualizations follow: firstly, that they are not themselves historical evidence but composite or derivatives of this evidence according to the HGIS schema employed and secondly, the sense that the visualized "output" is presented as authoritative—not least because a large number of complex procedural steps may have been undertaken in order to produce it. Even for historians accustomed to working with visual materials, this methodologically derived authority is likely to appear questionable. Such concerns may ultimately turn out to be ill-founded. Archaeologists are accustomed to referring to surveys made by other archaeologists which can be held up to critical scrutiny but are not considered inauthentic on the basis of being, effectively, composite images; similarly, there is no necessary reason why a visualization in HGIS should not be considered highly contingent in the context of ongoing research. Unfortunately, the exploratory or experimental potential of HGIS analysis is rarely apparent to non-specialists because this is not what GIS packages are principally designed to do. The historian's concern for authenticity in HGIS therefore stems mainly from the sense that the all important dialogue between the historian and their source materials is interrupted by the authority granted to visualizations as "outputs". Through a subtle process of translation and reduction, the resulting visualization rather than the source material becomes the research object.

The third difficulty pertains to *anachronism* as a consequence of decontextualization of source materials. Again, Lilley is right when he notes that there "remains reticence among historians on the uses of GIS, especially over fears over inputting often patchy or unrelated data and deriving from it spurious or anachronistic outputs and results" (Lilley 2012, 130). Anachronism has much to do with the requirement to reduce the complexity of source material and rely on visualizations that are themselves derivative or composites of the evidence. Again, the historian's intuitive concern is that the dialogue between the sources has been mediated—possibly in ways he or she does not fully understand—and that this process of mediation might go unreflected upon and therefore unarticulated.

Expert practitioners of HGIS can, of course, respond that these three concerns could be adequately addressed within an appropriate collaborative research environment—in a sense this could always be said to be the case. However, even if the practical point is ceded, it is still likely to be case that such epistemological concerns comprise an impediment on the development of such research collaboration. This is not an argument for the uniqueness of historical epistemology but does describe a tension between historical knowledge as it is conventionally achieved by historians working with sources and how this process might be affected by working in an HGIS. The emphasis on the HGIS visualization as the authoritative product of a well-defined methodological process not only raises question of reduction, authenticity and anachronism but also serves to erode the open-ended contingency that historical explanation generally constitutes. Ultimately, there is a danger that such visualizations "escape" the contextualization of the historical narrative.

11.5 HGIS as the "Creation of Phenomena"?

Given this state of affairs, a historian might ask himself or herself that while "yes, HGIS does perform well in managing my data and producing high-quality visualizations much faster than I could do so by hand—do I *really* need to use it, given my concerns about reduction, authenticity and anachronism?" To look at this question another way, can a contemporary social historian, for example, of urban or rural landscapes, armed with a GIS aspire to know more about the "world we have lost" than figures who had no such technology at their disposal such as Hoskins, Beresford or Dyos as a consequence?[6] There is little point in historians making the large investment in cross-disciplinary knowledge that HGIS entails to simply, in effect, replicate the outcomes of their existing de facto methods, especially given the epistemological issues that have been canvassed here. On the other hand, if historians thought that a commitment to HGIS would help them to develop new and distinctive perspectives on their sources, then this would clearly make it more worthwhile.

Given a due theoretical and methodological apprenticeship being served—it is arguably the capacity of HGIS to perform for areas of historical study what Hacking (1983) calls the "creation of phenomena" that is of most interest from this perspective. The phenomena of concern are particularly those regarded by Hillier (1996) as "non-discursive", in the sense that they do not typically lend themselves to description in the traditional language of historical discourse. Hacking argues that scientists traditionally saw their role as explaining or discovering phenomena presented by nature (i.e. "natural" science), whereas in fact, he argues that what scientists do is to *intervene* in nature to discover regularities through a system of experiment that entails the creation of phenomena. Contrary to conventional belief, he argues that experiments are rarely repeated (though they might be for pedagogical purposes), more often they are improved upon in order to test the original hypothesis. Hacking's broader argument is that the design of the experiment involves a process of "representing" the phenomena and then "intervening" in order to create new phenomena as part of a more open-ended process than the conventional notion of "experiment" implies.

Such an argument in relation to historical method may be controversial. Certainly, the relationship of historians to their sources is as phenomena presented rather than "created". Gaddis (2002) has argued convincingly that in proceeding from many causes to a few highly contingent consequences, historians generally acknowledge the impossibility of seeking positivistic "proof" of their accounts of causation equivalent to "experiments". Instead, Gaddis draws parallels with non-laboratory sciences such as geology and astronomy where variables cannot be controlled and experiments cannot be rerun. It is possible to sustain Gaddis' important distinction while still allowing Hacking's insight that refining knowledge is likely to involve, at some stage, some kind of iterative "experimental work" in which phenomena are created. At a push, we might allow this in the (admittedly ill-defined) sense of "thought experiments" in which historical source materials are critically consulted and various possibilities canvassed in advance of the construction of the historical account.

If the naïve view of historians has them looking to the sources to reconstruct the past "as it actually was", then a naïve view of science as "revealing" nature does not

[6] Hoskins (1955), Beresford (1967) and Dyos (1961); see also Ell (2011).

acknowledge the role of the scientist in the construction of scientific knowledge.[7] In rejecting these naïve positions historical knowledge can be characterized as emerging, if not through an experimental process as such, then through a heuristic-hermeneutic one—as the historian engages in a prolonged critical analysis of his or her (typically textual) source material. If historians would be unsympathetic to the idea that they are "creating phenomena"—nevertheless, they must concede that with their accounts, they are creating something (we might think of this as constituting "discursive phenomena"[8])! This is not, of course, to assert that such accounts are essentially fictitious (i.e. inauthentic)—rather that the practising historian, in dialogue with his or her source materials, knows that the account could have developed in "different ways" and that there is a process involved in deciding which is the right one. Somewhere in this heuristic-hermeneutic process emerges the "historical method" that involves the historian's sources in an ongoing dialogue with the sum of his or her historical knowledge in an intellectual process that is highly resistant to external mediation.

Yet where the phenomenon being addressed is one that does not lend itself to being revealed by traditional historical methods—at least not in the first instance—it is reasonable to propose that the historian should usefully call upon and/or develop alternative approaches. The description of the historical spaces of everyday spatial practice, Lefebvre's "perceived space" is surely such a case. Hillier and Hanson (1984, 43–45) present "recovering" such descriptions as fundamental aspects of being embodied in spatio-temporal reality, but it is not necessary to take such a cognitively reductive approach to history as Hillier sometimes appears to, in order to address this question. It is certainly not suggested that HGIS provides a ready-made description retrieval tool—there are no "black box" formulas. However, it is possible to envisage HGIS as providing an essential methodological component of a reinvigorated theory of historical spatial practice—approaching historical space as the nonrepresentational medium through which practice (action) indeed becomes *social* (Netto 2008). Such an empirically grounded approach to the built environment should facilitate a more nuanced understanding of the relationship between Lefebvre's triad of spatial practice, representations of space and representational spaces.

Approaching HGIS as a technology able to facilitate the creation of descriptive phenomena (the attendant risks of reductionism, authenticity and anachronism notwithstanding) should be regarded as legitimate by historians so long as it does not valorize the visualization as "end product" but rather treats visualization as a phase in an iterative process of historical thinking. Epistemologically, we might find that it is not be so different after all from what historians have been doing all along. The notion of heuristics is, arguably, one that historians and scientists can share. In this context, it refers to an open-ended "experimental" process of assessing the value of data sources to a given research endeavour from all perspectives, in the knowledge

[7] The chapter by von Lünen raises some epistemological issues involved in historical "reconstruction".

[8] My thanks to Charles Travis for this useful formulation.

that this process guarantees no positive outcome or "closure". If such an approach to historical research in a GIS environment can help focus scholarly attention on the description and interpretation of historical spaces, then this is surely a worth undertaking. Of course, it is one thing to outline a way forward, quite another for this to be translated into a method with any practical value for historical research—the author's own initial steps in this direction will be introduced in the following section. Clearly, much theoretical, methodological and empirical work remains to be done in order to overcome the disciplinary boundaries of HGIS, human geography and mainstream history in order to give any such method genuine substance.

11.6 Opening the Dialogue with Historical Sources in HGIS

Rather than conceiving of the visualization manufactured in an HGIS as possessing an inferior epistemological status to archival "source material", might it be possible to conceive of it in other ways? In an oft-quoted passage, the cultural historian Darnton explained how the historian interpreted historical texts: "passing from text to context and back again until he has cleared a way through a foreign mental world" (Darnton 1984, 6).

Can HGIS visualizations ever be accommodated to a similarly dialogic process for the description of historical space, or are the epistemological barriers of reduction, authenticity and anachronism simply too large to bear? Such a question relates to the broader challenge of a "topographical hermeneutics" posed by Sombart[9] but with a particular emphasis on the non-discursive, embodied spaces of practice. Certainly, there is a need for historians of the quotidian to reflect on their own research methods in order to identify those spaces of practice that resist description and representation in other domains. An interesting, though non-HGIS-based example of research that began to address such questions is the *Materializing Sheffield* project (McDonald 2006). One aspect of this involved architecture students constructing a detailed physical model of the late Victorian City using a wide range of sources but relying heavily on Ordnance Survey maps that "open up hitherto unknown worlds by revealing the sizes and relations of things", (Jones 2006). Although the emphasis here was very much on the built environment itself, it is clear that the participants felt they were able to comment on the "life lived" in the historic city. The significance of the method may lie less in the model-making per se than in the project's commitment to discovering new ways of "seeing" (in this case even "touching") the spatiality of a past built environment in ways that are difficult to achieve with traditional historical sources. Given adequate preparation, historians are surely close enough to their subject for their representation in HGIS (or other techniques) not to distort historical understanding but rather to "clear the way" for a consideration of historical spaces to become part of the interpretative dialogue.

[9]See von Lünen's chapter for a discussion of Sombart's idea.

The epistemological challenge is to "open up" the visualizations produced in a GIS to a process of translation in which they are approached less as would-be authoritative methodological "outputs" but rather as distinctive contributions to the hermeneutical process through which a wide range historical source materials are translated into historical narratives.

It has been argued that HGIS specialists need to reconsider the tendency of visualizations to emerge as "end products" in order to open up their method to the kind of ongoing dialogue that historians require in order to feel in control of the process and not lose sight of their source materials. Practically, this process should at least involve historians in the development of metadata schemas and encourage critical reflection on data standardization and visualization techniques. It should resist any attempt to assign generalizable validity to representations that must serve (in the first instance) the particular historical account, rather than render the historical account invisible as one among many "produced" by the method. An HGIS research environment sufficiently responsive to interdisciplinary dialogue can begin to address substantive concerns about the stifling of historical temporality by an inevitable emphasis on synchronic properties and the necessary practice of presenting "snapshots" that may suppress description of underlying processes. Ultimately, it is not the responsibility of the HGIS to "do time-space" but the responsibility of the historian to ask the right questions. The point of such research should be to open up new areas of enquiry, rather than simply streamline old ones in which the impact of HGIS is explained as much by institutional-logistical requirements than by research reasons.

That is not to say that the systematic use of HGIS by a proficient specialist, even in a fairly routine way, does not start to demand new kinds of descriptions of the built (inhabited) environment and raise new questions, where these are adequately realized. The ability simply to rapidly toggle between cartographic representations of a place from different historical periods, to zoom in and out at different scales and to query against a range of associated data, from land uses to more qualitative sources such as oral histories, permits a spatially sensitive interrogation of source materials to begin. Urban morphologists in the Conzenian tradition (Conzen 1960) have become accustomed to tracing the movements of plot boundaries and other features of the urban tissue, but the challenge of meaningfully relating these geographical descriptions to human activity is one that falls to historians—and there has been limited progress to date. Interestingly, it has taken one of the leaders of the Conzenian field to point this out. Reflecting on his review of current perspectives on urban morphology, M.P. Conzen made this argument in a passage worth quoting at length:

> So far, most of the approaches reviewed occasion little essential concern with history. To be sure, time, as an abstract dimension ($t_1, t_2, \ldots, t_n$), plays a role in quite a few of these perspectives, directly or indirectly, but not in the sense that is connected to the actual changing cultural experience of regions dear to the historian or geographer. Strangely, most mainstream historians do not take much interest in the built environment except for periods of documentary scarcity, concerned as they principally are with the actions and states of minds of people. (Conzen 2010, 57)

Greater familiarity with HGIS may itself generate sufficient phenomena (i.e. create new "ways of seeing") to enable built-environment historians more scope in exploring patterns of change and continuity over time. This is an endeavour already familiar to archaeologists of the contemporary past and material cultures (e.g. Mayne and Murray 2001) but is one that might inform research in social history more widely.

The author's own contribution to ongoing research at UCL into the historical built environment of suburban London has suggested the value of using historical maps in GIS as a useful starting point for questioning the dominance of powerful cultural representations (or "texts") of the suburbs that take little account of suburban space itself.[10] The existence of such texts (e.g. the suburb as "utopia" or "dystopia") makes it harder to conceive of the suburban as a particular kind of historically dynamic environment, productive of spatial practices that may beg questions of those frequently stereotyped representations that are often projected onto the suburb from without (Vaughan et al. 2009). Similarly, an HGIS analysis of the emergence of historical suburban road networks serves to highlight the fluidity and historical contingency of place names in relation to the complexity of "patterns on the ground" (Griffiths et al. 2010). In general terms, such work suggests how conventional language such as "road", "suburb" or "city" implies a purely synchronic description that, left unexamined, captures little of the historical specificity or temporal inscription of the spaces of practice to which they refer.

This perspective has been developed to a greater extent in research on the historical spaces of the cities of the British "industrial revolution"—particularly Sheffield (Griffiths 2008). Although initially developed outside an HGIS environment (and relying heavily on tracing paper as a consequence), this work strongly points to the potential value of this method in teasing out descriptions of those historical spaces of practice that are otherwise relatively inaccessible to decoding as "texts" in a manner analogous to language. The useful application of Hillier and Hanson's methodology of spatial description "space syntax" to this problem suggests the importance of the HGIS environment if such formal analytical techniques of spatial analysis (syntactical or otherwise) are to be adequately contextualized and brought into dialogue with other historical source materials. Where such contextualization does not take place, it is more likely that these approaches will appear reductively mechanistic or narrowly cognitive—revealing little that is valuable to the historian. This is an area where a good deal more research is needed, and the author's own work in this field is still at the experimental stage (Griffiths 2008, 2010, 2011a,b, 2012). So far, it has largely involved attempting to elucidate what Lefebvre might call the translation of "structures into textures"—or indeed "texts" into "textures" (Lefebvre 1991 [1974], 132). This is essentially a historicizing method in which the synchronic aspects of spatial representation—here given in terms of space syntax descriptions

[10]See "Towards Successful Suburban Town Centres" Project, UCL. Funded by the Engineering and Physical Sciences Research Council, Ref: EP/D06595X/1. http://gow.epsrc.ac.uk/NGBOViewGrant.aspx?GrantRef=EP/D06595X/1, last accessed 22 Apr 2012.

of eighteenth- and nineteenth-century Sheffield—are re-rendered diachronically and revealed as incomplete, open-ended and ambiguous with respect to their affordances for social life in specific historical contexts. The intention is to develop "spatial history" such that it becomes easier to conceive of and articulate the historical relationship between spatial practice itself and the emergence of symbolically privileged representational spaces—and to better understand how these in turn both contest and are informed by hegemonic spatial representations. For such work to succeed, it must contribute to historical scholarship in general rather than be defined principally by its association with particular methods.

11.7 Conclusion

This chapter has raised a number of epistemological difficulties that traditional, text-based modes of historical research and explanation encounter when using an HGIS environment. These include deeply entrenched disciplinary assumptions and preferences for addressing questions of "space", "place" and "location" that recur not just between history and HGIS but between GIS and geography more general. In using an HGIS, historians are likely to raise the issues of reductionism, authenticity and anachronism that have to do with their concern to preserve the primacy of their dialogue with the sources. It has also been noted that there is something of an irony in the fact that the "spatial turn" has, in some respects, distracted historians from an empirical consideration of space, particularly difficult to describe "everyday spaces". Yet, a reluctance to deal with more formal methodologies enabling spatial description may render a fuller understanding of the historical space of the quotidian spaces of practice, less likely. Where historians are able to suspend their concerns about authenticity, they may—it was speculated—find that HGIS provides a way into this complex issue by facilitating the description of spatial forms in time that would not have been possible otherwise. This is not simply to assert that such descriptions are somehow immanent and waiting to be "revealed" but rather to suggest how the HGIS may be reconceived as a vehicle for the "creation of phenomena" in the form of inauthentic (i.e. manufactured) visualizations that enable spatial descriptions to be hypothesized and their value explored as part of a research process involving the translation of diverse source materials into a historical narrative. However, it was also cautioned that the creation of such phenomena must occur in the context of a properly developed and rigorous historical epistemology of historical space in GIS. The well-defined methodological approach to GIS typically set out in textbooks needs to be changed to explicitly accommodate the "messiness" of historical source material and ensure that the methodological focus is not on the visualization as end product but as part of an ongoing heuristic-hermeneutic process in which a particular line of historical enquiry may be developed but possibly not reach resolution. Developing a similar argument, von Lünen in this book argues that a GIS can function as a "scratchpad" enabling the historian to participate in experimental, exploratory work that is not restricted by an undue concern for cartographic protocols. However, whereas von Lünen emphasizes the potential value of GIS to historians as one of *encoding* signs, translating the signified (map) to signifiers (clues, symbols), this chapter has focused rather on its potential to foreground the need for *descriptions* of spaces of practice within the signified spaces of the map itself. In so doing, it forces the question "how and to what extent do such space they *afford* signification and how should be described?" Following Lefebvre, it is crucial that such questions should be regarded as historical and not physical or transcendental in nature.

Continued improvement to the accessibility and usability of GIS—still a very specialist technique—and more broadly of GIScience as an emerging field of critical academic enquiry will hopefully facilitate such developments in practice so that HGIS becomes no more of a methodological barrier to historians than the use of a spreadsheet or the act of translating between languages. HGIS is not simply a "tool" for crunching large amounts of historical data but offers a pathway to "seeing" ("imagining"?) and thinking critically about places in the past. The aspiration is for "thick" descriptions of historical space based on archive research in which the dynamics of Lefebvre's perceived, conceived and lived space can better understood, particularly such that the semiotic content of space does not become divorced from the social dynamics of spatial practice. This in turn should improve understanding of the fundamental interrelationship of space, place and location in historical explanation. The development of such an epistemology would also advance historians' ability to engage positively with the discourse of spatial assemblages, networks and materializations that are emerging from archaeology, anthropology and geography, not least in considering how time and space are assembled from one another. HGIS, of course, cannot in itself provide such an epistemology, but it can provide a valuable theoretical and methodological locus for its development.

References

Baker ARH (2003) Geography and history: bridging the divide. Cambridge University Press, Cambridge

Bentley M (2006) Past and presence: revisiting historical ontology. Hist Theory 45:349–361

Beresford M (1967) New towns of the middle ages: town plantation in England, Wales and Gascony. Lutterworth, London

Biernacki R, Jordan J (2002) The place of space in the study of the social. In: Joyce P (ed) The social in question: new bearings in history and the social sciences. Routledge, London/New York, pp 133–150

Bourdieu P (1990) The logic of practice. Stanford University Press, Stanford

Braudel F (1976) The Mediterranean and the Mediterranean world in the age of Philip II, vol 2. Harper & Row, New York/London. Translated by S. Reynolds

Braudel F (1980) On history. Weidenfeld and Nicolson, London. Translated by S. Matthews

Cannadine D (1977) Victorian cities how different? Soc Hist 2:114–146

Conzen MRG (1960) Alnwick, Northumberland: a study in town-plan analysis. Institute of British Geographers, Publication 27, George Philip, London

Conzen MP (2010) The elusive common denominator in understanding urban form. Urban Morphol 14(1):55–58
Corfield P (1990) Walking the city streets: the urban Odyssey in eighteenth-century England. J Urban Hist 16(2):132–174
Darnton R (1984) The great cat massacre and other episodes in French cultural history. Penguin, London
Dyos H (1961) Victorian suburb: a study of the growth of Camberwell. Leicester University Press, Leicester
Ell P (2011) Local history in the twenty-first century: information communication technology, e-resources, grid computing, Web 2.0 and a new paradigm. In: Dyer C, Hopper A, Lord E, Tringham N (eds) New directions in local history since Hoskins. University of Hertforshire Press, Hatfield, pp 247–266
Gaddis JL (2002) The landscape of history. How historians map the past. Oxford University Press, Oxford
Gardiner P (1961) The nature of historical explanation. Oxford University Press, Oxford
Gibson J (1979) The ecological approach to visual perception. Houghton Mifflin, Boston/London
Giddens A (1984) The constitution of society: outline of the theory of structuration. Oxford University Press, Oxford
Gieryn T (2000) A space for place in sociology. Ann Rev Sociol 26:463–496
Gregory IS, Ell PS (2007) Historical GIS: technologies, methodologies and scholarship. Cambridge Studies in Historical Geography, Cambridge University Press, Cambridge
Griffiths S (2008) Historical space and the interpretation of urban transformation: the spatiality of social and cultural change in Sheffield c1770–1910. Unpublished Ph.D. thesis, University of London, London
Griffiths S (2010) Spatial transformation and urban culture in Sheffield c.1750–1830: patterns of change and continuity in civic ceremonial and popular processions. Unpublished paper presented at annual British society of eighteenth century studies 39th conference, St Hugh's College, Oxford, 5–7 January 2010
Griffiths S (2011a) Temporality in Hillier and Hanson's theory of spatial description: some implications of historical research for space syntax. J Space Syntax 1(2):73–96
Griffiths S (2011b) The use of space syntax in historical research: current practice and future possibilities. In: Proceedings of the eighth international space syntax symposium, Santiago, pp 1–26. http://www.sss8.cl/media/upload/paginas/seccion/8193_1.pdf. Accessed 14 Aug 2012
Griffiths S (2012) Public funeral processions and community identity in nineteenth-century Sheffield: the eighteenth-century as a period of transition between aristocratic and civic traditions. Unpublished paper presented to the annual conference of the social history society, University of Brighton 3–5 April 2012
Griffiths S, Jones C, Vaughan L, Haklay M (2010) The persistence of suburban centres in Greater London: combining Conzenian and space syntax approaches. Urban Morphol 14(2):85–99
Guildi J (2011) The roads to power. Harvard University Press, Cambridge, MA/London
Gunn S (2000) The public culture of the Victorian middle class. Ritual and authority and the english industrial city 1840–1914. Manchester University Press, Manchester/New York
Gunn S, Faire L (eds) (2012) Research methods for history. Edinburgh University Press, Edinburgh
Gunn S, Morris R (eds) (2001) Identities in space: contested terrains in the western city since 1850. Ashgate, Aldershot
Hacking I (1983) Representing and intervening: introductory topics in the philosophy of natural science. Cambridge University Press, Cambridge
Harvey D (1973) Social justice and the city. Edward Arnold, London
Hillier B (1996) Space is the machine. Cambridge University Press, Cambridge
Hillier B (2008) Space and spatiality: what the built environment needs from social theory. Constr Res Inf 36(3):216–230
Hillier B, Hanson J (1984) The social logic of space. Cambridge University Press, Cambridge
Hillier B, Leaman A (1973) The man-environment paradigm and its paradoxes. Archit Des 43:507–511

Hillier B, Vaughan L (2007) The city as one thing. Prog Plan 67(3):205–230
Hoskins W (1955) The making of the english landscape. Hodder and Stoughton, London
Jerram L (2011) Streetlife: the untold history of Europe's twentieth century. Oxford University Press, Oxford
Jones B (2006) Recreating nineteenth century Sheffield: The Sheffield urban study project. http://www.hrionline.ac.uk/matshef/pbj/MSpbj.htm. Last accessed 31 Mar 2012
Joyce P (2003) The rule of the freedom: liberalism and the modern city. Verso, London/New York
Kingston R (2010) Mind over matter? History and the spatial turn. Cult Soc Hist 7(1):111–121
Lefebvre H (1991 [1974]) The production of space. Carlton & Blackwell, Malden/Oxford
Lilley K (2012) GIS, spatial technologies and digital mapping. In: Gunn S, Faire L (eds) Research methods for history. Edinburgh University Press, Edinburgh, pp 121–140
Massey D (2005) For space. Sage, London
Mayne A, Murray T (eds) (2001) The Archaeology of urban landscapes: explorations in slumland. Cambridge University Press, Cambridge
McDonald S (2006) Materializing Sheffield. http://www.hrionline.ac.uk/matshef/index.html. Last accessed 31 Mar 2012
Netto V (2008) Practice, space, and the duality of meaning. Environ Plan D Soc Space 26(2):359–379
Pooley C (2000) Patterns on the ground: urban form, residential structure and the social construction of space. In: Daunton M (ed) The Cambridge urban history of Britain, vol 3, 1840–1950. Cambridge University Press, Cambridge, pp 429–465
Shoemaker R (2004) The London mob: violence and disorder in eighteenth-century England. Continuum, London
Thrift N (2008) Non-representational theory: space, politics, affect. Routledge, Abingdon
Tilley C (1994) A phenomenology of landscape: places, paths and monuments. Berg, Oxford
Tuan Y (1978) Space, time, place: a humanistic frame. In: Carlstein T, Parkes D, Thrift N (eds) Making sense of time. Timing space and spacing time, vol 1. Wiley, Chichester, pp 7–20
Urry J (2007) Mobilities. Polity, Cambridge/Malden
Vaughan L, Griffiths S, Haklay M, Jones C (2009) Do the suburbs exist? Discovering complexity and specificity in suburban built form. Trans Inst Br Geograp 34(4):475–488

Chapter 12
GIS and History: Epistemologies, Reflections, and Considerations

Charles Travis

12.1 Introduction

> In March, the gypsies returned. This time they brought a telescope and a magnifying glass the size of a drum, which they exhibited as the latest discovery of the Jews of Amsterdam. They placed a gypsy woman at one end of the village and set up the telescope at the entrance of the tent. For the price of five reales, people could look into the telescope and see the gypsy woman an arm's length away. "Science has eliminated distance," Melquíades proclaimed. "In a short time, man will be able to see what is happening in any place in the world without leaving his own house."
>
> ... José Arcadio Buendía ... conceived the idea of using that invention as a weapon of war ... He sent it to the government ... for several years he waited for an answer. Finally tired of waiting he bemoaned to Melquíades the failure of his project and the gypsy gave him ... some Portugese maps and several instruments of navigation ... the astrolabe, the compass and the sextant. José Arcadio Buendía ... became an expert in the use and manipulation of his instruments, he conceived a notion of space that allowed him to navigate across unknown seas, to visit uninhabited territories, and to establish relations with splendid beings without having to leave his study.
>
> That was the period he acquired the habit of talking to himself and walking through the house without paying attention to anyone ... as Úrsula and the children broke their backs in the garden growing banana and caladium ... He spent several days as if he were bewitched. Finally one Tuesday in December, at lunchtime, all at once he ... revealed his discovery to them:
>
> "The earth is round, like an orange."
>
> Úrsula lost her patience. "If you have to go crazy please go crazy all by yourself!" she shouted. "But don't try to put your gypsy ideas into the heads of the children."
>
> (Márquez 1967, 2–4).

From a geographer's point of view, one wonders (with some mischief), if historians who have recently embraced GIS methodologies are seen in a similar manner by their brethren, as José Arcadio Buendía, the nineteenth century Colombian patriarch

C. Travis (✉)
Trinity Long Room Hub, Trinity College, University of Dublin, Dublin, Ireland
e-mail: ctravis@tcd.ie

A. von Lünen and C. Travis (eds.), *History and GIS: Epistemologies, Considerations and Reflections*, DOI 10.1007/978-94-007-5009-8_12,

of Márquez's *One Hundred Years of Solitude*, who calculates centuries later what Eratosthenes first discovered in 240 BC. Disciplinary logos such as "spatial-humanities," "geo-humanities," and "spatial history" have emerged it seems, from "100 years" of academic "solitude" to mimic methodologies and technologies which human geographers have engaged in over the last half-century. While it is true that for over the past 2,000 years innovations in technology have facilitated Western spatial perceptions and practices, the digital revolution of the twenty-first century has profoundly reconfigured the scholarship potential of the sciences, the arts, and the humanities. New research frontiers have certainly been inaugurated by the internet and the diffusion of personal computers into the academic sphere. Immanuel Kant (1724–1804) observed that "geography and history fill up the entire circumference of our perceptions; geography that of space, history that of time," (Johnston et al. 2000, 410) and situated their disciplinary architectures as cornerstones in his conception of the modern university. But with the rise of our contemporary "corporate university" model, have these two disciplines become estranged from each other? By parsing the word "technology" from its Greek roots τεχνη (téchnē), "art, skill, craft," and λογια (logía), meaning "study of" we can reframe GIS scholarship to reconsider the "special relationship" between history and geography. Perhaps by re-conceptualizing GIS scholarship as one possible bridge between these two Kantian cornerstone disciplines, we can attempt to chart, navigate and survey the *terrae incognitaes* emerging in the wakes of the "spatial" and "cultural" turns which have influenced historical and geographical study over the past 100 years.

In some instances, because of its recent ubiquity, GIS has been misconstrued as a proxy for the Western geographical tradition, which historically, has been rather protean in nature. Simply stated, GIS is not "geography in a box." Like the Trojan horse set before the gates of Troy, geospatial software has been invited into the house of Herodotus. There it sits, seemingly as the heir to an ancient academic practice; instead, soldiers of a Euclidian epistemology infiltrate the courtyards of history, destroying the hermeneutical palimpsests of time and place, so tenderly cultivated by Herodotus' modern disciples. Alarmingly, scholars in an attempt to survive the throes of the digital revolution reconfiguring the "corporate university" either establish a division of labor between "humanists" and "technicians," or attempt to manipulate software programs, and flout algorithms, only to find themselves in the cul-de-sac of informatics, without grasping how critical spatial thought can illuminate their fields of study. Denis Wood, John Fels, and John Krygier contend that "using Google Earth may feel like magic, but it's not, or it's the magic of a Fred Astaire dance, effortless only because so long rehearsed" (Wood et al. 2010, 17), and at a more complex level, proprietary GIS software, if not employed within the context of the geographical tradition in which it was established, can foster a similar, albeit more sophisticated, and perhaps sinister illusion.

In *The Landscape of Time*, John Lewis Gaddis links the "ancient and ubiquitous practice" of map making with the "three-part conception of time," (past, present, future) employed by many historians (Gaddis 2002, 32). Gaddis notes that both disciplines "reduce the infinitely complex to a finite, manageable frame of reference" by imposing "artificial grids—hours and days, longitude and latitude—on temporal and spatial landscapes" and concludes "history, like cartography, is necessarily a *representation* of reality. It's not reality itself; indeed, if truth be told it's a pitiful approximation of a reality" (Gaddis 2002, 32).

However, as Edward S. Casey argues, perhaps the idea of mapping "needs to be liberated from its alliance with modern cartography so that it can resume its original sense of charting one's way into a given place or region" (Casey 2007, 512). Following Casey, we need to liberate geography's various perspectives from GIS and illustrate that the latter is a tool to serve the practices of the former. Once accomplished, GIS can be re-conceptualized as a tool to provide scholars with the ability to "write" spatial histories.

12.2 Origins of Western Geographical Practice

> Writing has nothing to do with signifying. It has to do with surveying, mapping, even realms that are yet to come.
>
> (Deleuze and Guattari 1988)

Contemporary Western geographic ontologies and epistemologies offer various perspectives. The roots of "geography"—translated from the Greek for "earth-writing"—share a common soil with the origins of history as a discipline: "for ancient Greeks, and to a lesser degree for Romans [...] *geographia*" represented "a literary genre more than a branch of physical science." Herodotus (484–425 BC) and Strabo (63 BC–24 AD) "sifted through a vast storehouse of traveler's tales in order to separate fact from fiction" and "retold those which they thought credible enough to claim a reader's attention" (Romm 1992, 3–4). In doing so, they contributed to establishing geography as an idiographic discipline anchored in ethics and politics. In contrast, Eratosthenes (276–195 BC) and Ptolemy (90–68 AD) employed mathematics to calculate the "spherical" nature of the earth and establish early map projections, in order to define geography as type of "geometry with names," foregrounding the discipline's later attempts to establish itself as a nomothetic practice. Initially "the geographer's science and storyteller's art [...] could not be fully detached from each other" (Olsson 2007, 32), provoking a clash between the perspectives of these two early schools:

> An academic controversy was waged over the reliability of geographical data in Homer's Odyssey. Strabo, who believed the Odyssey to be authentic and reliable, in a long and controversial passage leveled criticism against Eratosthenes for holding that Homer should be read as a poet and not as a scientific authority. (Wright 1996, 11)

From the time of this seminal argument, debates between the poetic and the positivistic have marked the evolution of Western geographical thought and practice. Michel de Certeau (1925–1986) observed that medieval and early modern map-makers conveyed spatial stories and histories by mapping religious pilgrimages and crusades "marked out by footprints" and "by pictures of successive events that took place in the course of the journey (meals, battles, crossing of rivers, or mountains)." He noted that "between the fifteenth and seventeenth centuries, the map became more autonomous" performing the function of a "theater." Map images were collated "on the same plane heterogeneous place, some *received* from a tradition and others *produced* by observation" (de Certeau 1988 [1984], 120–121). Brian Harley (1932–1991) states that from the seventeenth century onward, practices in Western cartography have "promoted a standard scientific model of knowledge and cognition" to "produce a 'correct' relational model of the terrain" being mapped (Harley 1989, 4). In this period, "two scopic regimes" began to "dominate European arts and sciences." One involved the development of Cartesian projection, allowing three-dimensional space to be represented on a flat surface in two dimensions; the other was the development of "perspectivism" which promoted the "grid" as an "organizing principle for all visual representation and seeing" (Pickles 2006, 80). In 1425, Filippo Brunelleschi (1337–1446), a Florentine artisan performed a "magical" experiment "near a doorway in a cobbled cathedral piazza" illustrating the optical illusion of the "vanishing point." This innovative technique had "irreversible implications for the entire future of Western art" (Pickles 2006, 84–85), and David Harvey notes that "perspectivism" which blossomed during the European Renaissance influenced early modern cartographical practices and "shaped the way of seeing for four centuries" (Harvey 1989, 244–245).

The science of optics during the Renaissance and the Enlightenment facilitated the ability of artists and scientists to represent perceptions "as in some sense 'truthful' compared to the superimposed truth of mythology and religion" (Harvey 1989, 244–245). According to John Pickles, such developments contributed to creating the modern "cartographic gaze." An "observer epistemology" posits a "universalist logic" anchored by "particular forms of parametric space, geometry, and scale." Describing it as a "controlling gaze", Pickles argues that this type of perspective frames "broad swathes of worldly complexity and enormity in miniature form for a discrete purpose" (Pickles 2006, 80). Consequently, the early modern "map theater" anticipated GIS by comprising "a totalizing stage on which elements of diverse origin were brought together to form the tableau of a 'state' of geographical knowledge and the perspective it offered" (de Certeau 1988 [1984], 121). In comparison, as R. J. Mayhew notes, during the Enlightenment, geography constituted "a distinct body of knowledge rather than a discipline" (Mayhew 1997, 7). Drawing from a diverse array of fields, interests, and insights, "both geography and literature were far more inclusive and more permeable categories in the eighteenth century, and their flexibility frequently drew them together [...] so individuals often considered as eighteenth century geographers frequently had wider interests and careers in writing comparable to individuals we now consider 'literary'" (Mayhew 1997, 43). In contrast, cartography, as Denis Cosgrove (1948–2008) observes, became a "profession that embraced such scientific practices as classification, quantification, and instrumentation to secure the truth of its visual records and representations." Cosgrove observes that states and nations soon established "surveying and cartographical institutions" to produce "topographical maps of their territory and of their colonial possessions." As a result, "statistical mapping" was implemented as a "significant tool of bureaucracy and social regulation, and map use and interpretation" became "widely taught in schools" (Cosgrove 2005, 37). Such practices influenced the development of geographical perspectives which coalesced in the late nineteenth century with the fallacy of environmental determinism.

12.3 Toward Postmodern Cartographies and Geographies

> ... the shape of this space is that of a river: not the surveyor's river which is simply a gap on the map, a frozen interval, but the river as serpentine motion, as an evolving pattern of vortices, expanding and collapsing. (Carter 1992)

From various quarters, "end of geography" debates greeted the dawn of the twentieth century. If the "globe" had indeed been mapped and enumerated in its entirety, it was argued, what was the purpose of the discipline? Despite these misgivings, geographical research on culture and place began to cut their own epistemological beds. In time, these streams of thought would coalesce to deconstruct the imperial and empirical cartographical "world" gazes of the fin de siècle. Witnessing the springtides of continental modernity eroding the cultural sandcastles of Edwardian and late Victorian England, Virginia Woolf (1882–1941) remarked that in 1910, or there about, "human character changed." Extrapolating Woolf's comment, Henri Lefebvre (1901–1991) argued that at the turn of the century, "the main reference systems of social practice in Europe disintegrated and even collapsed." Lefebvre noted that "what had seemed established for good during the belle époque of the bourgeoisie came to an end: in particular, space and time, their representation and reality indissociably linked." The shift in spatial perception and representation rippled first through intellectual and artistic spheres: "The old Euclidean and Newtonian space gave way to Einsteinian relativity. But at the same time, as is evident from the painting of the period—Cézanne first of all, then analytical Cubism—perceptible space and perspective disintegrated. The line of horizon, optical meeting point of parallel lines, disappeared from paintings" (Lefebvre 2005, 46). In the early twentieth century, the concept of "landscape" was being developed as a methodological lens by Carl Ortwin Sauer (1889–1975) and the Berkeley School of Cultural Geography. Sauer placed an emphasis on the temporal dimension of this lens:

> We cannot form an idea of landscape except in terms of its time relations as well as it space relations. It is in continuous process of development, or of dissolution and replacement. (Sauer 1925, 36)

The school emphasized a synchronic approach to mapping historical, cultural, and physical landscapes. Though empirically oriented in its methodology, the school recognized the significant role that subjective perceptions and impressions played in experiencing the distinct "senses of place" rooted in particular regions and cultures. However, in terms of standard twentieth century cartographic practice, as Lefebvre describes, "'One' continued to live in Euclidean and Newtonian space while knowledge moved in the space of relativity" (Lefebvre 2005, 46). These parallel tracks of spatial perception and practice manifested into two distinct schools of geographic thought: the former giving birth to "spatial science" and the quantitative revolution of the 1950s and 1960s' emphasis on mathematical modeling and computing (which inaugurated the development of GIS) and the latter shaping late twentieth century cultural, humanistic, and postmodern perspectives in geography. This "second cultural turn" in human geography adopted a "text" metaphor for "reading" landscapes, conducting fieldwork, and framing social and cultural life (Pickles 2006, 54). Conversely, "mapping" emerged as metaphor in the arts and humanities as scholars in these disciplines began to show "a strong interest in the spatial and geographical dimensions of literary and cultural texts" (Thacker and Brooker 2005, 1). The cross-pollination of methodological metaphors between disciplines informed Harley's seminal observation that:

> "Text" is certainly a better metaphor for maps than the mirror of nature. Maps are a cultural text. By accepting their textuality we are able to embrace a number of different interpretative possibilities. Instead of just the transparency of clarity we can discover the pregnancy of the opaque. (Harley 1989, 7–8)

In many quarters, however, GIS (dominated by positivistic epistemologies) had become by the end of the twentieth century "the *sine qua non* for geographical analysis and research in government, business and academia" (Dobson 1993, 431). Computer cartography framed the earth as a "*geo-coded* world" continuously "coded, decoded and recoded" by cybernetic language systems (Pickles 2006, 5). Internet and visual broadcast mediums facilitated the unprecedented phenomena of digital globalization. Marshall MacLuhan's (1911–1980) observation in 1964, that "after more than a century of electric technology, we have extended our central nervous system itself in a global embrace, abolishing both space and time as far as our planet is concerned" was radically prescient (McLuhan 1987, 3–4). The ubiquitous use of computers, "smart" phones, and consumption of 24/7 mass media outlets has profoundly shaped the contemporary geographic perceptions of the general public. Manuel Castells has argued that "the geography of the new history will not be made, after all, of the separation between places and flows, but out of the interface between places and flows" (Castells 1999, 295). Contemporary human geographical practices engage space as a dynamic "lifeworld," and a "quasi-material construct" produced by social interaction. Rather than acting as a passive container, space is considered an active agent, imbricated with human activity and perception which shapes, produces, and reproduces place socially, politically, and economically.[1] Nigel Thrift has fashioned a current geographical perspective which

[1]Historically, ideas on the nature of space have been considered by Gottfried Wilhelm Leibniz (1646–1716), David Émile Durkheim (1858–1917), Martin Heidegger (1889–1976), Maurice Merleau-Ponty (1908–1961), and Henri Lefebvre (1901–1991). Geographers such as Gunnar

is "non-representational" and concerned with "the geography of what happens" (Thrift 2008, 2). As a theoretical and "experimental" approach, it pulls the "energy of the performing arts into the social sciences" in order to "crawl out to the edge of the cliff of the conceptual" (Thrift 2008, 12). "Non-representational" theory considers space dynamically constituted from "more than culture or body" (Thrift 2008, 18). It perceives that with "the intervention of software" (Thrift 2008, 89), the human body has become a "tool-being" (Thrift 2008, 10) which is in symbiosis with the "new kind of electronic background of the time-spaces [. . .] shaping the world" (Thrift 2008, 89). The current geographical concern with performativity, social, and dynamic space, harnessed to "automated" mapping functions, cyber linguistics, and Web 2.0 (social media) 3.0 (semantic, geo-social, and 3D visualization) platforms, can facilitate deeper GIS engagements with historical research. By integrating various empirical, social, discursive, and symbolic approaches with emerging qualitative GIS practices discussed in the next section, alternate methodologies for "coding" and "writing spatial histories" can be considered.

12.4 "Rebooting" GIS to Write Spatial History

> **The Wizard** (covering up with the curtain): The Great Oz has spoken. Pay no attention to that man behind the curtain. . . the. . . Great. . . er. . . Oz has spoken.
> **Dorothy** (pulling aside the curtain and reprimanding): Who are you?
> **The Wizard**: (stuttering) I, I, I am the Great and Powerful Wizard of Oz.
> **Dorothy**: You are! I don't believe you.
> **The Wizard**: I'm afraid it's true. There's no other Wizard except me.
> **Scarecrow**: You humbug.
> **Tin Man**: Yeah.
> **The Wizard**: Yes. That's exactly so. I'm a humbug.
> **Dorothy**: (scolding) Oh, you're a very bad man!
> **The Wizard** (sadly): Oh, no, my dear, I. . . I'm a very good man—I'm just a very bad Wizard.[2]

Michel De Certeau's "map theater" provides a metaphor to reflect upon the possibilities for reconceptualizing and reconfiguring GIS methodologies for history. As we peer into the digital "proscenium" of the computer screen, our gaze, like a theater patron's, is focused by visual techniques established in the Renaissance. The operations we perform in GIS are crafted, after Shakespeare, like the lines of a Cartesian *worldplay*. We must recognize that the entire "illusion" of GIS, is scripted and constitutes a "codescape" of computer language and commands. GIS is the visualization of a cyber-syntax, which like Western geographical and cartographical

Olsson, Anne Buttimer, Yi-Fu Tuan, Nigel Thrift, David Harvey, Doreen Massey, and Edward Soja among others have drawn on these philosophers to inform their geographical methodologies.

[2] *The Wizard of Oz* (1939) (PAL DVD, Warner Home Video, 2006) ASIN: B00005NMWB, run time: 101 min.

epistemology has been formed by the "ocular-centrism" of the male gaze. But now, as in the "City of Oz" the "situatedness" of GIS "wizardry" has been exposed. In light of this, "qualitative" and "space-time" approaches in GIS have been recently pioneered by Mei-Po Kwan, Mariana Pavloskaya, Megan Cope, Sarah Elwood, Natalie Schuurman, Menno-Jan Kraak, and Alexandra Koussoulakou among others.[3] As Eric Sheppard notes we must now "consider how GIS—a technology [...] based on Boolean logics that social theorist can be quick to criticize—has shown a remarkable flexibility in connecting with other kinds of spatial logics and spatial representations, creating qualitative, feminist and ethnographic" approaches (Merriman et al. 2012, 9).

One of the main issues with GIS is that many practitioners rely on secondary data sources. In many cases, this data has been collected within the framework of a positivistic methodology, quantified, and preformatted to operate with commensurate GIS techniques. In this regard, the aphorism "if a hammer is your only tool, all your problems are going to look like nails"[4] is quite apropos, especially in light of David J. Bodenhamer's observation that the qualitative nature of humanities data "pose, far greater epistemological and ontological issues that challenge" GIS "technology in a number of ways" (Bodenhamer 2007, 102).

However, Pavlovskaya asserts that by engaging in "the ontological power of mapping," GIS can "be imaginatively expanded with qualitative research" and brought "outside [...] positivist epistemologies". She contends that "overall, spatial analysis in GIS today is largely qualitative, visual, and intuitive despite its insistent labeling as a quantitative method" (Pavlovskaya 2006, 2011). Indeed, a good portion of GIS attribute data is already qualitative in nature, including "names" (such as of owners of land parcels, businesses, or street addresses) or "types" (e.g., of roads, settlements, soils). Noting that such data is "not generally suitable for quantitative analysis," Pavlovskaya points out that it can "be queried and logically manipulated" by employing "the SQL (structured query language)" feature of GIS. She contends that this type of "querying reveals geographic features with a particular combination of attributes or locational characteristics." She further maintains that "even complex attribute and spatial queries require logical thinking and *spatial imagination* [emphasis mine] rather than statistical or mathematical skills" (Pavlovskaya 2006, 2013).

Ian Gregory has stated that GIS should be considered as a "data base visualizer,"[5] and this provides insight into considering how database construction and coding, coupled with the SQL and scripting capabilities of GIS, can harness what Bodenhamer argues is a technology that "offers the potential for an open, unique post-modern scholarship, and alternate construction of history and culture that embraces multiplicity, simultaneity, complexity and subjectivity" (Bodenhamer

[3]For a comprehensive discussion, see Cope and Elwood (2009).

[4]I attribute this to my mentor and colleague Professor Emeritus Richard Scott of Rowan University.

[5]Ian Gregory, lecture notes, Geospatial Methods for Humanities Research workshop, Digital Humanities Observatory Summer School, Royal Irish Academy, Trinity College, Summer 2010.

2007, 102). Whether one is engaged in quantitative or qualitative research, it is worth keeping in mind Pickles' observation that:

> . . . one of the most difficult lessons for anyone to learn is the way in which their own worlds are geographically coded; to understand the relationship between the visible and invisible, the proximate and the distant, and to recognize the complex folds of past and present that constitute place and experience as we know it. (Pickles 2006, 81)

Bodenhamer notes that qualitative tropes engaged by the humanists include "variability, interdependency, contingency, uncertainty, agency, and nuance," which seemingly, are resistant to "any method or tool that appears to reduce complex events to simple schemes" (Bodenhamer 2007, 102). In light of this, LaDona Knigge, and Meghan Cope's recent discussions of "grounded theory" as a methodology which operates at concrete and abstract levels of analysis have relevance for developing GIS approaches which engage the qualitative tropes espoused by historians. Its aim is to build "theories" from data about the social worlds "grounded" in people's everyday experiences and actions. Knigge and Cope note that "grounded theory's" methodology is iterative, reflexive, and inductive; it involves "multiple stages of collecting, coding, categorizing, and analyzing data to enable themes to emerge through iterations of 'constant comparison'" (Knigge and Cope 2006, 2024).

Bodenhamer suggests that by "locating historical and cultural exegesis more explicitly in space and time," GIS can identify "patterns," facilitate "comparisons," enhance "perspectives," and illustrate "data" (Bodenhamer 2007, 107). Furthermore, John Corrigan states that "data like butterflies become something else when they are 'captured' and pinned to a foamcore board." He proposes that a "humanities GIS chases the butterfly rather than nets and pins it" (Corrigan 2010, 82). "Grounded theory" provides a research strategy for scholars engaged in GIS to "chase the butterfly" and identify categories, patterns, internal consistencies, or inconsistencies, as well as correlations and disjunctions, in their data (Knigge and Cope 2006, 2024). If we extrapolate "grounded theory" to the historian's realm, its methodology provides a means to think about how GIS can perform the "topographical hermeneutics" as discussed in Alexander von Lünen's chapter. The next section will illustrate how such an approach was used in GIS to "write a spatial history."

12.5 GIS and the Phenomenal Cartographies of Time-Space

> "Do you know the phrase *watershed moment*, buddy?"
> . . . It was one of those annoying linguistic shortcuts that show up on cable TV news shows, day in and day out. Others include *connect the dots* and *at this point in time.*
> [. . .]
> "Do you know where it comes from? The origin?"
> "Nope."
> "Cartography. A watershed is an area of land, usually mountains or forests, that drains into a river. History is also a river . . . Sometimes the events that change history are widespread – like heavy prolonged rains over an entire watershed that can send a river out of its banks.

> But rivers can flood even on sunny days. All it takes is a heavy, prolonged downpour in *one small area* of the watershed. There are flash floods of history, too. Want some examples? How about 9/11? Or what about Bush beating Gore in 2000?" (King 2011, 57)

Since the publications of Paul Vidal de la Blache's *Tableau de la Géographie de la Françe* (1903) and Carl Sauer's *The Morphology of Landscape* (1926), the scales of time and space have respectively infiltrated the disciplines of geography and history. In the 1970s, geographer Torsten Hägerstrand (1916–2004) employed "space-time paths" to map how individuals navigate their environments. Place was represented cartographically on a standard two-dimensional (x and y) horizontal plane, with location and destination plotted as zero-dimensional points. Time was then represented graphically on a vertical axis (z), allowing Hägerstrand to create a three-dimensional "aquarium" to illustrate a specific "cube" of space-time. John May, Nigel Thrift, and David Harvey have since explored the influence of "time-space" and "time-space compression" on human geographies. Edward Soja has applied Lefebvre's notion of "rhythmanalysis" in urban geography. More recently, Mei-Po Kwan developed algorithmic techniques to chart Hägerstrand's "space-time paths" in GIS to study gender issues and accessibility in American cities. Doreen Massey outlines how time and space "co-implicate" and are "integral" to each other (Massey 2005), and Eric Sheppard states that "our debates in geography too often set time aside or position it in an orthogonal Newtonian third dimension [. . .] which freezes our way of thinking about the world." Sheppard argues that geographers "need to attend to history, to the irreversibly emergent nature of space-time as these complexes co-evolve" (Merriman et al. 2012, 9). Massey and Sheppard's observations echo Sauer's emphasis on the historical dimension of landscape. Conversely, historians have been considering the "emplacement" of time (Ethington 2007, 487). Jack Owens, long a proponent of GIS, has called for a "geographically integrated history" (Owens 2007). P.J. Ethington notes that the "all human action presumes locations in spacetime, which materializes each place" and argues that "the past cannot exist in time: only in space. Histories representing the past represent the place (topoi) of human action" (Ethington 2007, 487).

Thus, historical interpretation is reconceived as "the act of reading places, or *topoi*." Ethington argues that "knowledge of the past [. . .] is literally cartographic," (Ethington 2007, 466) and claims that "the incalculable volume of historical writing on all subjects should be thought of as a map because the past can only be known by placing it, and the way of knowing places is to map them" (Ethington 2007, 487). GIS provides a platform upon which to cartographically collate and visualize various layers of historical narrative. As David J. Staley notes, GIS allows "a kind of multidimensional emplotment: a single story organized from multiple and heterogeneous elements as a spatial totality" (Staley 2007, 43). Furthermore, Anne Kelly Knowles states that the use of GIS in history "compels writers to think graphically and forces spatial thinkers to come to grips with the subtlety of historical texts" (Knowles 2008, 3).

12.6 Writing Kavanagh's "Spatial History"

By engaging Mikhail M. Bakhtin's theory of "historical poetics" with space-time mapping techniques in GIS, I performed a "topographical hermeneutics" on the contrasts between the rural and urban *lifepaths* patterns of the Irish writer Patrick Kavanagh (1904–1967) and the images of place in his poetry.[6] During his childhood, adolescence, and young adulthood, Kavanagh lived in Inniskeen Parish, County Monaghan, where his family owned several fields and farmed the land, but in 1939, relocated to Dublin. His migration and poetry can be seen respectively as a metaphor and an articulation of the major rural to urban population shift which occurred in Ireland over the course of the twentieth century. Kavanagh began to write poetry in the 1920s, drawing inspiration from his local environment. However, two distinct "senses of place" can be seen to distinguish Kavanagh's depiction of Inniskeen in the *Ploughman and Other Poems* (1936) from its portrayal, 6 years later in the epic poem *The Great Hunger* (1942). The poems in the collection, for the most part, are infused with a "topophilic" and pastoral impression of the people and the drumlins of his native parish. The epic poem, in contrast, declares that the same landscape is experiencing an "apocalypse of clay" (Kavanagh 1942, 33). Written from the perspective of an urban milieu, it elicits a sense of "topophobia" verging on "topocide" or "the annihilation of place" (Porteous 1988). Despite Sheppard's reservations, the figurative use of orthogonal visualizations provides the means to conduct a "topographical hermeneutical" exploration comparing the daily life pattern Kavanagh established as a "farmer-poet" with the one he later maintained as an urban "man of letters." Conducting fieldwork in drumlin country Monaghan and the streetscapes of south Dublin, I employed Bakhthin's "historical poetics" to map the different spatial patterns distinguishing his separate rural and urban *lifepath* movements, with the change in his poetic voice, over time. Bakhtin notes that "chronotopes" (literally "time-spaces") are *leitmotifs* which serve "as the primary point from which 'scenes' unfold 'within narrative space' and act as "an optic for reading texts as x-rays of the forces at work in the culture system from which they spring" (Bakhtin 1986, 425–426). Consequently, chronotopes possess biographical dimensions, which intersect with their counterparts in literature to reflect the particular environments and periods experienced by the writer. Bakhtin asserts that "a definite and absolutely concrete locality serves as the starting point for the creative imagination" and concludes "this is a piece of human history, historical time condensed into space" (Bakhtin 1986, 49). I created a pair of corresponding rural and urban "time-space" (chronotope) visualizations plotted in three dimensions and coded with Kavanagh's biographical and literary data from the 1930s and 1940s.

Kavanagh's early rural voice inherited the "Gaelic bardic tradition of *dinnseanchas*" (knowledge of the lore of places; Kiberd 1997, 107). Situated in Inniskeen, his writing practices reflected an intimate "geography based on *seanchas*, in which there is no clear distinction between the general principles of topography

[6]For a more detailed discussion, see Travis (2010).

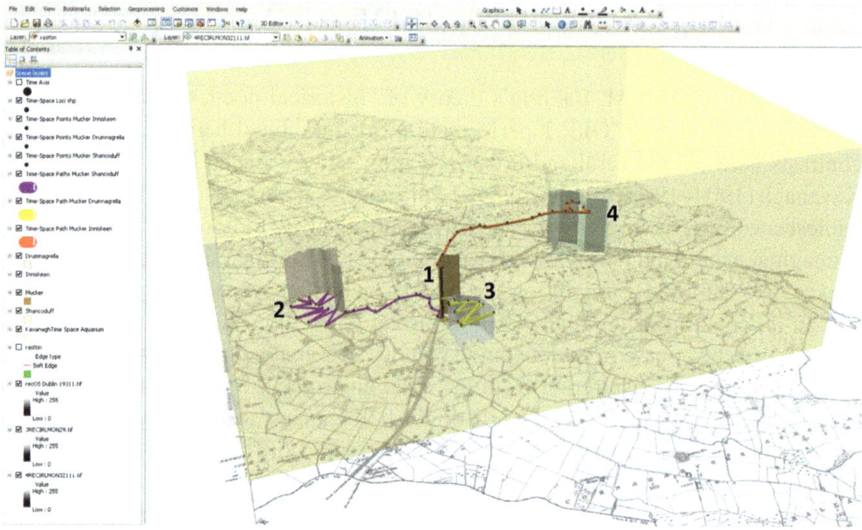

Fig. 12.1 "Time-space" aquarium Patrick Kavanagh lifepaths Inniskeen Parish (1920–1939) (© Charles Travis)

or direction-finding and the intimate knowledge of particular places" (Bowen 1975/1976, 115). This sensibility shapes the poetic imagery of his 1936 collection, and Kavanagh's literary and biographical narratives are rooted in Bakhtin's chronotope of the "idyll." This denotes "an organic fastening-down, a grafting of life and its events to a place, to a familiar territory, [...] the conjoining of human life with the life of nature, the unity of their rhythm [and] the common language used to describe phenomena of nature and the events of human life" (Bakhtin 1981, 225).

I created the base of the Inniskeen "time-space" visualization by digitizing two historical map (Irish Ordnance Survey (OS) layers 29 and 32, first published in 1911) layers. The GIS visualization in Fig. 12.1 contains the following sites whose Gaelic names were Anglicized by the original British OS in 1835: Kavanagh's native townland of Mucker (Fig. 12.1/No. 1), the fields of Shancoduff (*Seanchua Dubh* "Black Hollow"; Fig. 12.1/No. 2) and Drumnagrella (*Droin* "Ridge"; Fig. 12.1/No. 3), and the village of Inniskeen (*Inis Caoin* "Beautiful Island"; Fig. 12.1/No. 4). In the visualization, three polylines in ascending order trace Kavanagh's daily movements and form a composite *lifepath* pattern of his early years in Inniskeen Parish. The visualization was mapped from biographical information and fieldwork conducted in the parish. This consisted of walking and driving surveys, visiting sites, and "tracking" paths and routes which Kavanagh might have taken over the terrain of the parish on a daily basis. Though the routes mapped are speculative, they are also pragmatic in the sense that the topography of the townlands has changed very little, if at all. The shortest contemporary pedestrian routes seem consistent with an observation that Kavanagh once made: "My idea of a cultural parochial entity was the distance a man would walk in a day in any direction. The center was usually the place where oneself lived though not always." (Kavanagh 1986, 243)

As the visualization illustrates, rising before dawn, Kavanagh would have performed chores around his family's house in Mucker. Then having the energy from a full night's rest, he probably would have set out to work in the farthest four fields of Shancoduff (Fig. 12.1, No. 2, violet polyline) located on the northern slope and basin of a sizable drumlin. Fieldwork surveys noted that it took approximately 15–20 min to walk from Mucker to Shancoduff, a circuitous route, given the formidable elevation of the drumlin. Depending on the day and season, Kavanagh would have spent different amounts of time working in the fields. Perhaps he would eat lunch on site, or return to Mucker. It was speculated that generally in the afternoon, he would have gone to the fields at Drumnagrella, adjacent to his family's holding, after expending the morning's energy on the family's farthest holding in Shancoduff (Fig. 12.1, No. 3, yellow polyline). In the evening, he might have travelled to Inniskeen village to socialize (Fig. 12.1, No. 4, red polyline). The three respective polylines which trace Kavanagh's speculative movements during the day are anchored by his family's home at Mucker. In the visualization, a composite pattern of centrality emerges, linking this site by the routes he may have walked to the family's fields and village of Inniskeen. Kavanagh's depiction of his birthplace and other townlands in the poem *Inniskeen Road: July Evening* (1935):

> The bicycles go by in twos and threes-
> There's a dance in Billy Brennan's barn to-night,
> And there's the half-talk code of mysteries
> And the wink-and-elbow language of delight
> [...]
> A road, a mile of kingdom. I am king
> Of banks and stones and every blooming thing,
> (Kavanagh 1972, 19)

is laced with lyricism, dry humor, laconic observation, and suffused with a mystic-like illumination of the parish's surrounding vista and its idiosyncratic social geography. However, *The Great Hunger* published in 1942, deconstructs the centered spatiality of Kavanagh's earlier life as a farmer-poet and reveals a topophilic vision of Inniskeen Parish:

> He stands in the doorway of his house
> A ragged sculpture of the wind
> October creaks the rotted mattress
> The bedpost fall. No hope. No. No lust.
> The hungry fiend
> Screams the apocalypse of clay
> In every corner of this land.
> (Kavanagh 1942, 33)

Juxtaposing the GIS "time-space" visualizations of Kavanagh's composite rural and urban daily lifepaths, one is struck by their contrasting spatial patterns; the former is centrifugal, while the latter is centripetal. "Idyllic" images associated with the poems he composed in a rural environment are transformed, and the chronotope of "peripherality" which marks a space "lost in a cyclical, natural, or static time warp [and a place] forgotten by history bypassed by history" begins to shape his poetic image of Inniskeen parish (Leersson 1994, 4). The GIS visualization in Fig. 12.2

Fig. 12.2 "Time-space" aquarium Patrick Kavanagh lifepaths Dublin (1939–1942) (© Charles Travis)

records the fieldwork exercise used to identify sites in Dublin which Kavanagh was known to frequent in the early 1940s. His movements through the streets of the city were reconstructed by plotting polylines in a similar ascending manner. Socialized as a farmer to rise early, Kavanagh would write in his bedsit—and frequent the Grand Canal (Fig. 12.2/No. 1) in the morning time. The black polyline in the upper right hand corner of the GIS visualization captures the speculative movements of his morning routines. In the afternoon, Kavanagh would walk to meet his brother for lunch at a Westland Row cafe (Fig. 12.2/No. 2) and afterward wander into a "cultural sphere" which contained the National Library on Kildare Street (Fig. 12.2/No. 4), Hanna's Bookstore on the corner of Dawson and Nassau Streets (Fig. 12.2/No. 4), and Bewley's Cafe on Grafton Street (Fig. 12.2/No. 6). The yellow polyline charts these midday and afternoon movements. His afternoon and evening perambulations could overlap with stints in McDaid's Public House on Harry Street (Fig. 12.2/No. 5) and the Palace Bar on Fleet Street (Fig. 12.2/No. 7). Located across Westmoreland Street from the old *Irish Times* offices, the Palace Bar was as "A Café Literaire" (Connolly 1942, 36). Its space approximates Bakhtin's chronotope of the "salon and parlor", where "webs of intrigue are spun, denouements occur and finally [...] dialogues happen" (Bakhtin 1981, 246). In Bakhtin's "Historical Poetics," the Palace Bar served as "a barometer of political and business life" and was a place where "political, business, social, and literary reputations" were "made and destroyed and careers begun and wrecked." Bakhtin notes "the graphically visible markers [...] of biographical and every day time are concentrated and condensed," in such a space, and the Palace Bar provided Kavanagh entry into Dublin's literary circles. Nicknaming it the "Malice Bar," Kavanagh recalled that Bertie Smyllie "the giant Hemingway-esque editor of the *Irish Times*" (Bakhtin 1981, 247) instituted a nightly branch office of the *Times* at the bar: "almost everyone who counted in journalism and the arts was to be seen in the Palace Bar at some time on the evening of the week: F.R. Higgins, poet and Abbey Theatre director, M.J. MacManus, novelist and literary editor of the *Irish Press*" (Quinn 2001, 125). As a result, Kavanagh secured writing commissions and entered into Dublin's literary circles. The vibrancy of the bar can be juxtaposed against the insularity of the rural pub depicted in *The Great Hunger*:

> The frosted townland of the night.
> Eleven o'clock and still the game
> Goes on and the players seem to be
> Drunk in an Orient opium den.
> (Kavanagh 1942, 24)

After the Palace Bar closing time of 11 p.m., Kavanagh would return to the bed-sit which he shared with his brother (Fig. 12.2, the red polyline). By employing GIS techniques in conjunction with Bakhtin's "historical poetics," Kavanagh's literary and biographical landscapes become "not only 'graphically visible' in space but also 'narratively visible' in time" (Folch-Serra 1990, 258). Kavanagh's spatial performances in Inniskeen Parish (Fig. 12.3) convey such a perspective; once a centering "idyllic" pattern as captured in his poem, *Ploughman* (1930):

> I turn the lea-green down
> Gaily now,
> And paint the meadow brown
> With my plough.
> (Kavanagh 2005, 1)

When juxtaposed against his streetscape perambulations in Dublin (Fig. 12.3), the pattern reverts to the "cyclical" spatiality of an insular, closed system, represented in *The Great Hunger*:

> But the peasant in his little acres is tied
> To a mother's womb by the wind-toughened navel-cord
> Like a goat tethered to a stump of a tree-
> He circles around and around wondering why it should be.
> (Kavanagh 1942, 29)

The rural landscape in the GIS "time-space" visualizations of Inniskeen Parish implies centrifugal movements in space, which appears rooted in a locus, but ultimately leads nowhere. In contrast, his Dublin streetscape peregrinations illustrate a centripetal tendency which in a rural environment imply the decline, ruin, and entropy of a society. Kavanagh intimated as much when he wrote in his semi-autobiographical novel *The Green Fool* (1938):

> When I arrived in Mucker the natives were beginning to lose faith in the old, beautiful things. The ghost of a culture haunted the snub-nosed hills. (Kavanagh 1939, 11)

However, in Dublin's urban milieu, Kavanagh's centripetal movement led to interactions with dynamic nodes of culture and linkages with different networks

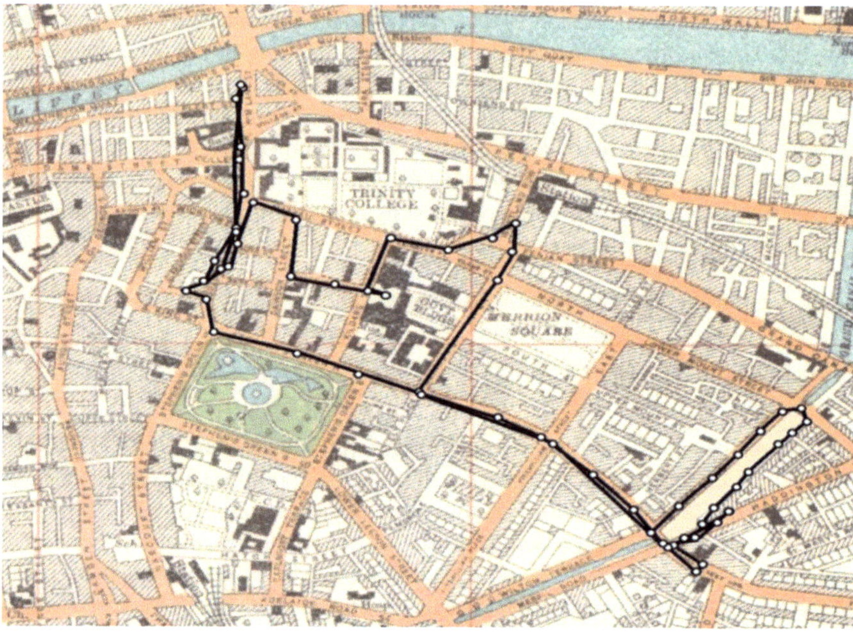

Fig. 12.3 Patrick Kavanagh lifepaths Dublin (1939–1942) (© Charles Travis)

of knowledge and new forms of thought and style. We can see that a core space was created by the collective dynamics of his urban pedestrian trajectories (Figs. 12.4 and 12.3). Geographer Edward Soja has coined the word *synekism* to convey the "stimulus of urban agglomeration" (Soja 2000, xv) which this "particular spatial dynamic" imparts "around a dominant and centripetal core." Soja notes that *synekism* provides the "context for active and affective processes of [...] innovation, development, growth, and change" (Soja 2000, 13). Such a particular spatial awareness informs Kavanagh's poem, *I Had A Future* (1952), a bitter-sweet reflection on his early days in Dublin:

> ...Gods of the imagination bring back to life
> The personality of those streets,
> Not any streets
> But the streets of nineteen forty.
> [...]
> It is summer and the eerie beat
> Of madness in Europe trembles the
> Wings of the butterflies along the canal.[7]

The GIS "time-space" visualizations of Inniskeen and Dublin "chase the butterflies" of Kavanagh's changing poetic perceptions of place. The visualizations

[7] P. Kavanagh, "I Had a Future" (1952) in Kavanagh (2005, 261–262).

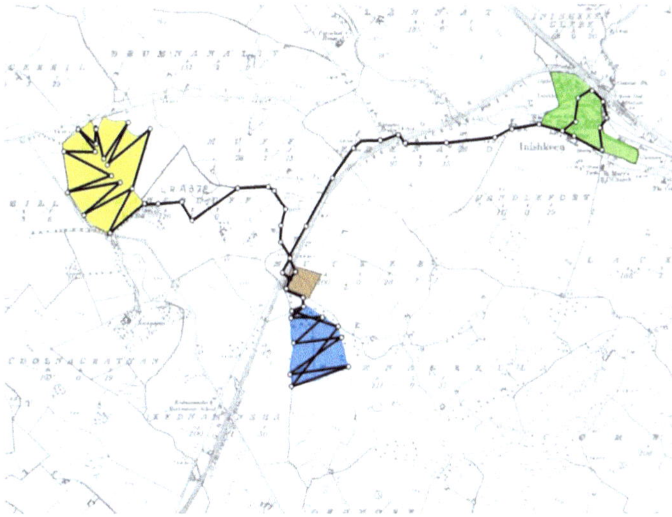

Fig. 12.4 Patrick Kavanagh lifepaths Inniskeen Parish (1920–1939) (© Charles Travis)

suggest that a *synekitic* shift in Kavanagh's spatial movement and environment contributed to a "chronotropic" shift in the literary perspectives with which he framed subsequent landscape depictions of his native parish. The respective visualizations, though captured as static images for this chapter, were created in GIS upon a dynamic 3D platform. Drawing upon both gaming and filmic techniques, the visualizations were manipulated in GIS to show different angles, perspectives and scales.

The cartographer Bill Cartwright has proposed the metaphor of "The Gameplayer" to engage map user's to employ "gaming skills" to explore geographical information; this coincides with the plethora of computer games and strategies which could facilitate the creation of similar platforms and strategies for GIS use in historical studies. In addition, the ability to create GIS 3D animations can be informed by Denis Cosgrove's observation that "like driving and flying, the cinema offers a kinetic spatial experience characteristic of modernity, transforming the possibilities for representing space cartographically" (Cosgrove 2005, 50). Studies have shown that human visual acuity is triggered by movement and change. In addition, GIS can appropriate the technological medium of film with its techniques of flashback, jump cut, and ensemble story line. Historical streetscapes and landscapes can be framed cinematically. The ability of filmic techniques, as Walter Benjamin noted in 1936, can

> assure us of an immense and unexpected field of action. Our taverns and our metropolitan streets, our railroad stations and our factories appeared to have us locked up hopelessly. Then came film and burst this prison-world asunder by the dynamite of the tenth of a second, so that now, in the midst of its far-flung ruins and debris, we calmly and adventurously go travelling. With the close-up, space expands; with slow motion, movement is extended.
> (Benjamin 1992, 229)

By synchronizing Lefebvre and Benjamin's writings on history and place, an alternative perspective on considering GIS as platform to cinematically "emplace" dynamic historical narrative is possible. Lefebvre contends that "history emerges from insignificant tales [*du récit anecdotique*], annals, and epic poems" (Lefebvre 1970, 106), while Benjamin holds that "history is the subject of a structure whose *site* [emphasis mine] is not homogeneous, empty time, but time filled with the presence of the now [*Jetztzeit*]" (Benjamin 1992, 261). Benjamin further states, "I have long, indeed for years played with the idea of setting out the sphere of life—*bios*—graphically on a map" (Benjamin 2009, 295). GIS provides a platform upon which to engage such ideas. Staley notes that "when considering [such] two- and three-dimensional spaces like a map, we need to broaden our definition of narrative, or at the very least devise a definition for spatial narratives." He further observes that "symbols—words, lines, polygons colors—link together in a web, not a sequential chain" (Staley 2007, 41). Integrating Bahktin's idea of the chronotope in a dynamic 3D GIS mapping enmeshed with the coordinate grid of latitude and longitude, allows phenomenological impressions to synchronize with the Cartesian; as a result, a dialectical space which can be examined for further study and analysis emerges. As Mei-Po Kwan states,

> the representational possibilities of GIS can be used for enacting creative discursive tactics that disrupt the dualist understanding of geographical methods where visual images, words, and numbers are used together to compose contextualized cartographic narratives in geographical discourse. (Kwan 2002, 272)

The ability to iterate spatial scales with "gaming" and "filmic" techniques provides a powerful research tool which used in tandem with qualitative methods, and critical theory can be employed to research and write further three-dimensional spatial histories in GIS.

12.7 Conclusion

> [T]he heretic cartographer is finally prepared to move from the theory of art history to the practice of human geography. (Olsson 2007)

Denis Cosgrove has observed that "in the past decade, thinking in Science and Technology Studies has tended to dissolve epistemological distinctions between art and science." Just as science's claims to a universal truth have diminished in the face of postmodernity, "Modern artists have rejected aesthetics as the defining feature of their work" (Cosgrove 2005, 51). In the twenty-first century, academics operate (to paraphrase Donna Haraway) within a "cyborgscape." Nigel Thrift contends that such new "space-time backgrounds" emerging as a result of complex cybernetic systems and software language are producing a "qualculative world"[8] where we

---

[8]For a further discussion of "qualculativeness," please see Thrift (2008, 582, 604).

"literally and metaphorically 'write' space in different ways" (Merriman et al. 2012, 19). Tools such as GIS are merely extensions of the intellectual and cybernetic environments which have developed to date and in which we reside, work, and play. However, if "historical GIS" is to become viable and sustainable, we need to "reboot" the tool and harness the ontological and epistemological strengths of the humanities and arts—not vice versa. Pavlovskaya notes that because of its development and conventional use, it is a greater challenge "to represent theoretical relationships" and the "results of qualitative analysis in GIS" such as the "spatial configurations of networks, relationships, activities, meaning of places and events" to the "flows that link people and places" (Pavlovskaya 2006, 2013). My own personal approach to digital scholarship (geographical, historical, cultural, literary) has always been to "critically play" with the tools while keeping the qualitative tropes of arts and humanities firmly in mind.

One way I think about this is to consider how music is charted and performed. Melodies and rhythms are schematically diagrammed on the staves according to mathematical principles. However, when translated by a musician, these representations create sonic vibrations in space, which cross the threshold from the quantitative to the qualitative—creating an entire liminal space of performance and reception. In many ways, GIS is like my Fender Stratocaster. I believe that GIS scholarship will proliferate only by developing its own unique language, tools, and methodologies. As Corrigan states, using GIS in the humanities promotes the possibility that "three key referencing systems—space, time, and language—might be engineered in such a way that changes in one ripple into the others" (Corrigan 2010, 85). Epistemological shifts stemming from the digital revolution of the twenty-first century suggest that perhaps the spiral staircase of history is leading us to an ontological "plateau" or "plane" of perception, where we will rediscover as the ancient Greeks and Romans did before us, that the "geographer's science" and the "storyteller's art" of the historian cannot be fully detached from one another (Romm 1992, 3–4).

References

Bakhtin M (1981) The dialogic imagination: four essays. University of Texas Press, Austin

Bakhtin M (1986) Speech genres and other late essays. University of Texas Press, Austin

Benjamin W (1992) Illuminations. Fontana, London

Benjamin W (2009) One-way street and other writings. Penguin, New York

Bodenhamer D (2007) Creating a landscape of memory: the potential of a humanities GIS. Int J Humanit Arts Comput 1(2):97–110

Bowen C (1975/1976) A historical inventory of the *dindshenchas*. Studia Celtica 10/11:113–137

Carter P (1992) The sound in between: voice, space, performance. New South Wales University Press, Sydney

Casey E (2007) Boundary, place, and event in the spatiality of history. Rethink Hist 11(4):507–512

Castells M (1999) Grassrooting the space of flows. Urban Geogr 20(4):294–302

de Certeau M (1988 [1984]) Spatial stories. In: de Certeau M (ed) The practice of everyday life. University of California Press, Berkeley

Connolly C (1942) Horizon. V. Vanguard, Dublin
Cope M, Elwood S (eds) (2009) Qualitative GIS: a mixed method approach. SAGE, London
Corrigan J (2010) Qualitative GIS and emergent semantics. In: Bodenhamer DJ, Corrigan J, Harris TM (eds) The spatial humanities: GIS and the future of humanities scholarship. Indiana University Press, Bloomington, pp 76–88
Cosgrove D (2005) Maps, mapping, modernity: art and cartography in the twentieth century. Imago Mundi 57(1):35–54
Deleuze G, Guattari F (1988) A thousand plateaus. Continuum, New York
Dobson JE (1993) The geographical revolution: a retrospective on the age of automated geography. Prof Geogr 45:431–439
Ethington PJ (2007) Placing the past: 'Groundwork' for a spatial theory of history. Rethink Hist 11(4):465–493
Folch-Serra M (1990) Place, voice, space: Mikhail Bakhtin's dialogical landscape. Environ Plan D Soc Space 8(3):255–274
Gaddis JL (2002) The landscape of history: how historians map the past. Oxford University Press, Oxford
Harley J (1989) Deconstructing the map. Cartographica 26:1–20
Harvey D (1989) The condition of postmodernity: an enquiry into the origins of cultural change. Blackwell, Oxford
Johnston RJ, Gregory D, Pratt G, Watts M (eds) (2000) The dictionary of human geography. Blackwell, Malden
Kavanagh P (1939) The green fool. Penguin, New York. http://books.google.de/books?id=rBz-2F_ADCwC
Kavanagh P (1942) The great hunger. Cuala, Dublin
Kavanagh P (1972) Inniskeen road: july evening. In: Kavanagh P (ed) Collected poems. Penguin, Newbridge
Kavanagh P (1986) Patrick Kavanagh: man and poet. National Poetry Foundation, Orono
Kavanagh P (2005) Collected poems. Penguin Classics, Penguin Books, New York
Kiberd D (1997) Inventing Ireland. Vintage, London
King S (2011) 11/22/63. Scribner, New York
Knigge L, Cope M (2006) Grounded visualization: integrating the analysis of qualitative and quantitative data through grounded theory and visualization. Environ Plan A 38(11):2021–2037
Knowles AK (ed) (2008) Placing history: how maps, spatial data, and GIS are changing historical scholarship. ESRI, Redlands
Kwan MP (2002) Is GIS for women? Reflections on the critical discourse in the 1990s. Gend Place Cult 9(3):271–279
Leersson J (1994) The western mirage: on celtic chronotope in the European imagination. In: Collins T (ed) Decoding the landscape : contributions towards a synthesis of thinking in Irish studies on the landscape. Centre for Landscape Studies, Galway
Lefebvre H (1970) La révolucion urbaine. Gallimard, Paris
Lefebvre H (2005) Critique of everyday life, vol III: from modernity to modernism (towards a metaphilosophy of daily life). Verso, London
Massey D (2005) For space. SAGE, London
Mayhew RJ (1997) Geography and literature in historical context: Samuel Johnson and eighteenth-century english conceptions of geography. School of Geography, Oxford
McLuhan M (1987) Understanding media: the extensions of man. Ark, London
Merriman P, Jones M, Olsson G, Sheppard E, Thrift N, Tuan YF (2012) Space and spatiality in theory. Dialogues Hum Geogr 2(1):3–33
Márquez GG (1967) One hundred years of solitude. Harper & Row, New York
Olsson G (2007) Abysmal: a critique of cartographic reason. University of Chicago Press, Chicago
Owens J (2007) What Historians want from GIS. ArcNews 29:4–6
Pavlovskaya M (2006) Theorizing with GIS: a tool for critical geographies? Environ Plan A 38(11):2003–2020

Pickles J (2006) A history of spaces: cartographic reason, mapping and the geo-coded world. Routledge, London/New York
Porteous JD (1988) Topocide: the annihilation of place. In: Eyles J, Smith D (eds) Qualitative methods in human geography. Polity, Cambridge, pp 75–93
Quinn A (2001) Patrick Kavanagh: a biography. Gill & Macmillan, Dublin
Romm JS (1992) The edges of the earth in ancient thought: geography, exploration, and fiction. Princeton University Press, Princeton
Sauer CO (1925) The morphology of landscape. University of California Publications in Geography, Berkeley
Soja E (2000) Postmetropolis: critical studies of cities and regions. Blackwell, Oxford
Staley D (2007) Finding narratives of time and space. In: Sinton D, Lund J (eds) Understanding place: GIS and mapping across the curriculum. ESRI, Redlands
Thacker A, Brooker P (2005) Introduction: locating the modern. In: Thacker A, Brooker P (eds) Geographies of modernism: literatures, cultures, spaces. Routledge, London
Thrift N (2008) Non-representational theory: space/politics/affect. Routledge, London
Travis C (2010) Abstract machine—geographical information systems for literary and cultural studies: 'Mapping Kavanagh'. Int J Humanit Arts Comput 4(1/2):17–37
Wood D, Fels J, Krygier J (2010) Revisiting the power of maps. Guilford Press, New York
Wright JK (1996) Human nature in geography. Harvard University Press, Cambridge

Chapter 13
Maps of Change: A Brief History of the American Historical Atlas

Edward L. Ayers, Robert K. Nelson, and C. Scott Nesbit

13.1 Introduction

History and GIS pursue their shared fascination with finding patterns in complex bodies of evidence in quite different ways. History relies on narrative and discursive explanation, while a GIS relies on data and image. By putting history and GIS into conversation, we can emphasize what is distinctive, useful, and taken for granted in each. History's representation of space tends to be thin and under-examined; it can benefit from the depth and rigor of GIS. History, in turn, can help GIS be more attentive to mutability and interaction, helping maps tell stories.

The two perspectives can benefit each other because social change emerges from multiple structures—political, economic, religious, cultural, discursive, and ideological—connecting and conflicting in complex combinations. Because these structures interact in unpredictable ways, patterns of time are as uneven as the patterns of space. The union of history and GIS can help unravel those remarkably knotty processes as neither can alone.

Of course, long before the emergence of GIS, scholars integrated history and geography in their work. Historical accounts have used mapping, literally and figuratively, from the days of Herodotus, while geographic accounts often present history as necessary context. The historical atlas, which dates back to the sixteenth century, is the most notable genre that fused the two perspectives. The various ways historical atlases have told the story of the American experience provide particularly revealing perspectives on the interaction of history and geography. An

E.L. Ayers (✉)
University of Richmond, Richmond, USA
e-mail: eayers@richmond.edu

R.K. Nelson • C.S. Nesbit
Digital Scholarship Lab, University of Richmond, Richmond, USA
e-mail: rnelson2@richmond.edu;snesbit@richmond.edu

A. von Lünen and C. Travis (eds.), *History and GIS: Epistemologies, Considerations and Reflections*, DOI 10.1007/978-94-007-5009-8_13,

historical atlas—a collection of maps devoted to telling the story of a geographic area across time—is more than the sum of its individual images. It defines an intentional vision and interpretation of a complex subject, combining image and language, in addition to geography and history. The historical atlas, by putting maps in the context of time, necessarily creates a narrative, a story of how things have come to be in a particular place. The ways historical atlases have shown change over time and conveyed interpretations (or not) can inform the prospective value and contributions of GIS and the discipline of geography for historical scholarship.

13.2 The American Historical Atlas's Past

The chronicle of the atlas of American history is brief. While maps of all sorts have long portrayed the changing landscape of what is now the United States, Americans have only rarely paused to represent the patterns of their past in a comprehensive way. In the years before the Civil War, historical atlases were created mainly for use in the new nation's schoolrooms, offering a vision of the United States as one of inevitable and heroic growth. In the 1870s, the new *Statistical Atlas of the United States* built on the detailed information of the national census to offer powerful portrayals of a nation reunited after devastating civil war, changing with breakneck speed. That remarkable atlas, for all its innovations, represented history only as a backdrop to progress (Schulten 2011).

In 1896, Albert Bushnell Hart, an early champion of history as a profession, offered a blunt assessment in his *Guide to the Study of American History*: "historical maps abound, but they are for the most part inaccurate [...] There is a great need of an elaborate historical atlas of North America, worked out from the documentary source." Hart's devotion to primary evidence and expertise reflected the emergence of a new and newly professionalized discipline of history. His interest in mapping reflected a fascination among historians with the relationship between people and their landscape, a perspective made famous in 1893 by Frederick Jackson Turner's "The Significance of the Frontier in American History," one of the most influential works ever produced by the profession (Channing and Hart 1896, 53).

Hart's contemporary John Franklin Jameson took up the challenge of developing an expert historical atlas. The first graduate of the PhD program in history at the Johns Hopkins University in 1882, Jameson spent the bulk of his career assembling a disciplinary apparatus for history. Two years after defending his dissertation, he helped establish the American Historical Association; in 1895, he became the first editor of the *American Historical Review*. Jameson, instrumental in creating the National Archives, worked tirelessly to create robust and accessible collections of published primary sources and reference materials. Such projects, he felt, were "the necessary preliminaries" that would allow the still-maturing historical professoriate to produce "generations of successful individual work."[1]

A "first-rate atlas of American historical geography" quickly emerged as one of the most pressing of the "preliminaries" Jameson identified as the groundwork for sophisticated and rigorous historical research in the future. In 1902, Jameson proposed the idea of an historical atlas to the president of the newly formed Carnegie Institute in Washington, an organization dedicated to furthering scientific discovery. Only an institute like the Carnegie could support the "expensive amount of pioneer research work" and coordinate the "extensive cooperation" that the production of such an atlas would require. Then teaching at the University of Chicago, Jameson helped to plan the agenda for the Carnegie Institute's Bureau (later the Department) of Historical Research, and in 1905, he accepted a position as the bureau's director.[2]

Work on the atlas began in earnest in 1912 under the leadership of Charles O. Paullin. A historian of the navy, Paullin had studied and completed his doctorate at the University of Chicago during the same years that Jameson taught there. For the next 15 years, Paullin led the development of maps and accompanying text with substantial input and guidance from Jameson. In his preface to the *Atlas*, Paullin called it a "composite work." It was indeed a deeply collaborative project involving dozens of historians from universities and historical societies, archivists, librarians, and government officials. Particular historians were asked to help plan and research series of maps related to their areas of expertise; many of them drafted the text that accompanied the plates. Other scholars, archivists, and librarians provided more modest but still significant support by supplying the historical data for particular maps. In 1927, the Carnegie Institute decided to shut down the Department of Historical Research with the atlas still unfinished. Two years later, they enlisted the American Geographical Society (AGS) to bring "the enterprise to a close." The final product comprised 166 plates consisting of hundreds of maps, with 145 folio pages of explanatory and bibliographic text. John K. Wright, librarian at the AGS and a historian with a doctorate from Harvard, oversaw the completion of and wrote the introduction for the *Atlas*. When the volume was finally published in 1932, Paullin received credit as author and Wright as editor.

When the *Atlas of the Historical Geography of the United States* appeared, scholars acclaimed it a "monument to American scholarship" (cf. Wilgus 1932; Karpinski 1934). In 1933, the volume received the Loubat Prize, awarded by Columbia University every 5 years to the best work on the history, geography, or philology of North America. Appreciation of the *Atlas* has not diminished over time. Writing nearly a half-century after its publication, Lester J. Cappon simply observed that "no compilation subsequent to the Paullin and Wright *Atlas* has replaced it" (Cappon 1979). In a 1997 study of historical atlases, the British scholar Jeremy Black determined that the *Atlas* "was arguably the best hitherto in the world" (Black 1997, 120).

[1]Franklin to Daniel C. Gilman, 14 February 1902, in Donnan and Stock (1956, 80); also Higham (1965, 20–25).

[2]Franklin to Gilman, 14 February 1902, in Donnan and Stock (1956, 81).

Remarkably, a century after it was imagined, the *Atlas* remains unmatched in its comprehensive representation of American history, though a number of atlases have drawn inspiration from the 1932 work to good effect. Cappon's excellent *Atlas of Early American History* explicitly drew comparisons to Paullin's work. Atlases of particular aspects of US history, particularly *The Settling of North America*, by Helen Tanner and a team of historians and cartographers, have likewise brought care and sophistication to their subjects. Mark C. Carnes and John A. Garraty turned their keen eyes to the genre of the atlas as well, building on their successful collaboration on the *American Nation* textbook. Yet for nearly a century now, an atlas conceived in the early years of the historical profession in this country has found no successor in its scope, innovation, and ambition (cf. Cappon et al. 1976; Tanner et al. 1995; Carnes et al. 1996).

The impact and endurance of the *Atlas* is largely attributable to its intellectual coherence and ambition. The volume emerged from the progressive tradition of American historiography that focused on the role of the physical environment and economic interest in shaping the American nation and its regions. The *Atlas* linked the origins of the political system, religious institutions, language, and culture of the United States to Europe and more specifically to England, but Wright nevertheless insisted that "we are not European." The landscape of the North American continent had shaped American history so that everything inherited from the Old World had "undergone a subtle but profound transformation amid a new environment" (Wright 1932, 354).

The initial sections of the *Atlas* were devoted almost entirely to that "new" and transformative natural environment: maps of ocean currents, soil regions, mineral resources, forests, topography, temperature, and rainfall. That environment, the organization of the *Atlas* implied, gave fundamental shape to the agricultural, commercial, political, religious, and educational activities mapped in the latter half of the *Atlas*. For Paullin and Wright, American Indians seemed to exist in a state somewhere between "nature" and "culture"; they placed maps representing Indian lands early in the *Atlas* sandwiched between sections on European cartography and European exploration of the continent.

The editors of the *Atlas* depicted the formation of the American democratic character in constant interaction with the landscape. They aimed to represent the "facts of geography and of history that condition and explain the development of the United States" (Paullin 1932, xi). They framed most of the maps on a national scale and depicted most of the variations within as regional or sectional, highlighting the persistence of the region as a geographic and cultural unit. Sectional differences, Wright explained, were natural: "If the frontier has passed, sectionalism will remain [...], perhaps, as long as men live on this continent, for sectionalism is rooted both in human nature and in the nature of the land itself" (Wright 1932, 356). Natural and cultural phenomena interacted to produce history on the landscape, and the boundaries of this history were imprinted by deeply rooted regional cultures.

This interest in linking human and physical phenomena at the regional level captured a distinct historiographical moment. Wright and Paullin's interpretive strategy fit equally well with Frederick Jackson Turner's theories on national and sectional development and Charles and Mary Beard's explication of class and sectional interest in American history. For each of these historians, sections were the most illuminating geographic features in the shaping of the United States. Distinct topographical patterns laid the groundwork for sectional interest and antagonism.

For Turner, frontier and forest clearings ingrained themselves on the people of the United States by creating a self-reliant, restive national character. Those geographic features also created distinct sections and sectional interests. In time, the West emerged across the Alleghenies not simply as the "raw material for power" to be fought over by northern and southern states but as a self-conscious section with its own interests. The rise of the West, Turner believed, was a constitutional and geographical problem of the highest order to eastern interests. Agricultural interests held vast power in the Senate in proportion to their relative population and wealth. "On a map constructed so as to give to each state a space proportioned to its population, or to its income tax," Turner observed, "instead of to its dimensions in square miles, the Western lands would shrink in their map space in a startling fashion." The uneven sectional geographies and even the geology that underlies the sections had persistently affected US economics, politics, and culture. The result was a "vast and varied Union" of competing sectional interests, not unlike those envisioned by Paullin and Wright (Turner 1925, 260, 267–268, 280).

In Charles and Mary Beard's view, the emergence of sections marked the United States as fundamentally different from Europe. The new world distended social conflict over space. In Europe, labor and capital fought in close physical proximity, whereas in the United States, agricultural and manufacturing interests came into conflict as sections. In the nineteenth century, evidence of this conflict was nowhere more apparent than in votes on the tariff. Paullin and Wright agreed with the Beards' emphasis: the editors of the *Atlas* devoted five maps to votes on the tariff before the Civil War and another five to tariff votes afterward (Beard and Beard 1927, 663–680).

Though this historiographical perspective and agenda shaped the content and organization of the *Atlas*, Paullin and Wright explicitly foreswore interpretation. They insisted that the *Atlas* was not a history at all but instead a reference book. Each map was "a refinement [...] of the raw materials for historical research, comparable to a document carefully edited with textual criticism but without historical interpretation." They believed that "original studies will undoubtedly be based upon the data" presented in the *Atlas*, but they did not present the volume as an original argument. John Franklin Jameson modestly described his role fostering works like the *Atlas* as "making bricks without much idea of how the architects will use them." Charles Paullin pursued that ambitious but self-effacing role for the better part of two decades.[3]

The development of the *Atlas* was driven by this disciplinary agenda, not by any technical innovations in cartography or mapping more generally. The cartographic tools available to Paullin and Wright had changed little since the emergence of

[3]Paullin (1932, xi). Jameson to Henry Adams, 31 October 1910, in Donnan and Stock (1956, 136).

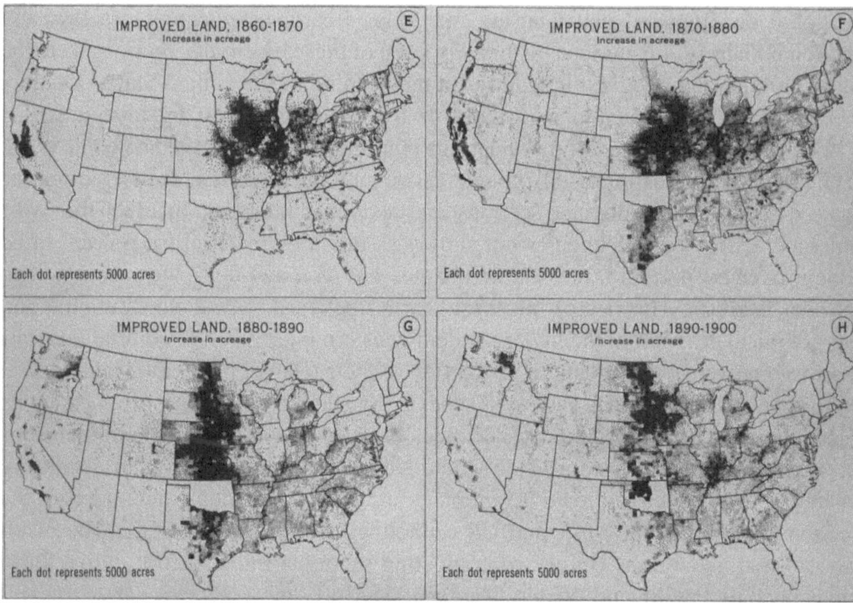

Fig. 13.1 Thematic maps from the *American atlas* showing land improvement in the USA between 1860 and 1900 (Taken from Paullin (1932))

photomechanical production techniques in the 1880s and did not change substantially over the 20 years they worked on the project (Cook 2002). The 1932 *Atlas*, drawn with exactitude and imagination, built on the cartographic traditions of the preceding centuries even as it subtly foreshadowed those that would follow.

Most interesting from our current perspective, the *Atlas* took on the fundamental challenge confronting an alliance of history and GIS: representing change. Paullin and Wright struggled to find effective ways to convey change over time given "the somewhat unadaptable medium of the map." They used three techniques. Their "simplest" device, as they put it, was to develop sequences of maps that spatially represented a variable at a number of specific moments—which presidential candidate received the most votes at the county level during each election cycle, for example, or the population density of census tracts each decade. Another device mapped relative change over a given period of time. On two series of maps, for example, they plotted increases and decreases in improved land during each decade, bringing "out in sharp relief the vital movements with which the historian is likely to be most immediately concerned" (Paullin 1932, xiv) (Fig. 13.1).

Paullin and Wright showed change, too, by plotting on the same map events that happened at different moments. Their use of this technique could be very simple, such as when they mapped changing boundaries between states. It could also be more complex, superimposing graphs on maps that showed change over time for one or more variables in multiple locations. Sometimes the maps struggled to contain the

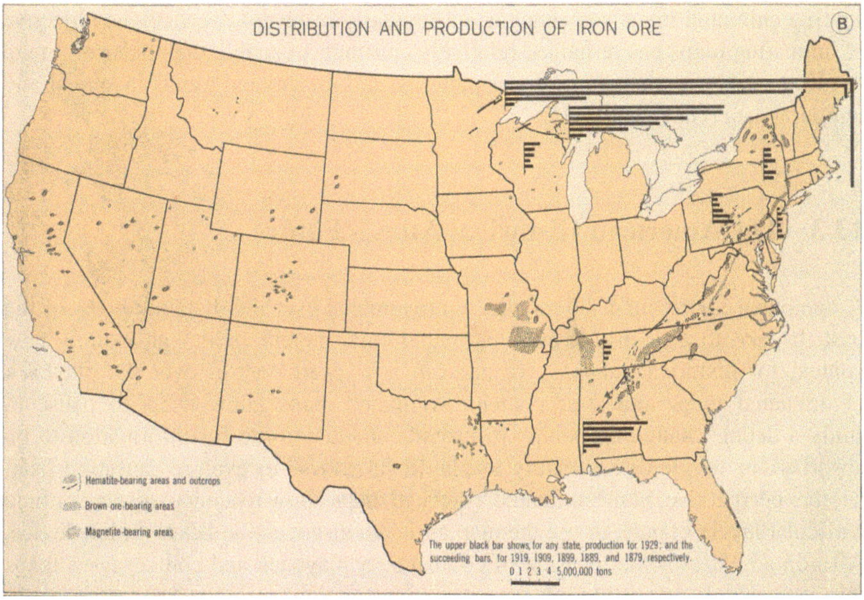

Fig. 13.2 Thematic maps from the *American atlas* showing the distribution and production of iron ore in the USA (Taken from Paullin (1932))

multiplicity, as when the *Atlas* displayed bar graphs showing the distribution and production of iron ore for each state at six intervals between 1879 and 1929. The x-axis represented tonnage and the y-axis time. When those horizontal bars collided with the eastern edge of the map, their orientation changed to travel vertically along the y-axis. As one reviewer complained, "Such statistics would have been better depicted in tables or graphs." Paullin and Wright, so successful otherwise, strained in this instance to represent complex and highly variegated change (Fig. 13.2).[4]

Even as they brought the *Atlas* before the public in the early 1930s, Wright recognized the possibilities of electronic media. He admitted that print imposed a "large problem" in showing historical change, and he imagined transcending its limits. "The ideal historical atlas," he dreamed, "might well be a collection of motion-picture maps, if these could be displayed on the pages of a book without the paraphernalia of projector, reel, and screen." The ideal historical atlas, in other words, would look much like animated maps available on the computer screens of the twenty-first century (Paullin 1932, xiv).

The allure of animated maps, the most obvious way to represent change over space, has continued to beckon since the days of John K. Wright's vision of the future of historical geography. Mark Harrower has traced the long history of animated cartography in some detail, in which celluloid film and then videotape preceded digital storage and distribution. While the means of producing and

[4]Paullin (1932, plate 6), and Merk (1933).

sharing animated maps has changed substantially in recent decades, the objective of animating maps has remained relatively constant: to enable the reader of a map to more easily see change, discover patterns, and understand trends in spatial and temporal data (Harrower 2004).

13.3 The American Historical Atlas's Future

A computer environment offers vast improvements over the dynamic maps of the past. In fact, that electronic and distributed environment creates an entirely new context for historical maps. A reader can now exert control over the playback of animated maps and their content. Animated maps allow users to pause to study a detail, change the tempo to provide adequate time for information to be absorbed, or toggle directionality to play it in reverse to explore causation from another perspective. Some animated historical maps allow readers to display or hide particular layers of data, to use the map as a data discovery and exploration device. Affording readers some measure of control over playback and content turns maps into increasingly powerful tools for active investigation.

So far, digital maps created as tools for investigation resemble Paullin and Wright's conception of their *Atlas* as an "aid to historians and to teachers and students of history" but not "history" itself. That is true for digital humanities in general. Digital archive projects, for instance, account for the great majority of digital history, certainly in terms of volume if one counts the Library of Congress, the National Archives, and projects presented by libraries and historical societies. Many of the ambitious early projects in the digital humanities—*the Valley of the Shadow*, the *Walt Whitman Archive*, and the *Rossetti Archive*—amassed substantial archives of historical and literary materials but saved analysis for print scholarship. We might think of this early digital scholarship much as John Franklin Jameson thought of his work, as space for potential interpretation, made available to broad audiences.[5]

Intellectual changes in the last decades have made it possible to envision a digital atlas as an important contribution to scholarship. Debates about GIS and positivism are only the most recent examples of a long-running argument among geographers whether maps are objective and scientific documents or subjective and argumentative documents. In the 1940s, John K. Wright emphasized the subjective element of maps and called on geography to reconnect with the humanities, championing claims at odds with his denial of interpretation in the *Atlas*. In "Map Makers Are Human," Wright emphasized that regardless of geographers' and cartographers' efforts to be objective and scientific, inevitably every map "gives a partly subjective interpretation" for an obvious reason: "the fact that maps are drawn by men and not turned out automatically by machines." That meant that maps

[5] A 2003 experimental article by William G. Thomas and Ayers, sponsored by the *American Historical Review*, did integrate evidence and interpretation, including geographic evidence, within the same digital entity (Thomas and Ayers 2003).

could be “influenced by human shortcomings” but also that they could reflect the “judgment, discrimination, critical acumen” of the geographer (Wright 1942, 527, 532, 541). In his presidential address to the Association of American Geographers in 1947, Wright made a more forceful claim, urging his colleagues to reestablish connections with the humanities, particularly history. While he did not dismiss the importance of “scientific geography,” he championed a humanities-inflected geography that he associated with “scholarship.”

> There is something to be said for considering scholarship, as distinguished from science alone, as our métier. All science should be scholarly, but not all scholarship can be rigorously scientific. Scholarship, moreover, embraces not only the natural sciences and social studies but also the humanities—the arts and letters—inquiring no less into the world of subjective experience and imaginative expression than into that of external reality.
>
> Reconnecting with history and the humanities, Wright proposed, would “render the study of geography more powerful than it would now seem to be in firing the artistic and poetic imaginations of students and public” (Wright 1947, 14–15).

With the advent of GIS beginning in the 1960s, Wright’s “obvious” fact that maps were inescapably subjective because they were “not turned out automatically by machines” not only was no longer obvious, it was no longer a fact. What Jerome E. Dobson labeled “Automated Geography”—computationally driven cartography and spatial analysis—unleashed passionate arguments over the epistemological foundations of the discipline of geography. Some geographers championed GIS as reenergizing a positivistic approach to geography, while others decried GIS as fostering a naive empiricism or worse. Most geographers, it seems, eventually adopted a pragmatic position between these extremes. They have recognized the practical value of GIS as a powerful set of tools but recognize that those tools do not lessen the centrality of interpretation in their work.[6]

Many historians, unaware of these debates, have been suspicious and critical of GIS as an intellectual hand-me-down from the social scientific, governmental, and corporate worlds, considering it an ill fit with the questions, methods, and assumptions of history and other humanities disciplines. Historians deal in ambiguity and surprise, they stress, aspects of social life for which GIS seems poorly equipped. Yet even as some have ignored or resisted using geospatial technologies, many historians have become increasingly attentive to “place” in their scholarship. Studies have proliferated detailing how colonial borderlands, nineteenth-century neighborhoods, postwar suburbs, and the globe itself have been historically constituted as places by men and women who imagined them and thereby brought them into being. In the 1970s and 1980s, theorists and historians began to investigate the ways that cultural power defined and was enacted in space. Cultural historians and geographers have found public, institutional, and domestic spaces to be expressions of cultural power and necessary stages for its deployment.

---

[6]Dobson (1983), Taylor (1990), Goodchild (1991, 335–337), and Wright et al. (1997).

D. W. Meinig's monumental four-volume interpretation of life on the North American continent, the most ambitious effort since Paullin and Wright to unite history and geography in this context, brought a humanistic geographer's perspective to this work in the last quarter century. In a series of brilliant meditations on the expansion of American people, government, and institutions across the vast landscape, Meinig addressed issues he found interesting and ignored others. He did not assume the responsibility for consistent coverage that the makers of an atlas typically shoulder. He abjured GIS in favor of carefully crafted illustrations of patterns, giving more space to text than to image, connecting to older traditions of cartography while offering bold and original interpretation (Meinig 1986–2004).

Even historians most interested in uncovering the profound and intricate spaces created by past societies have been hesitant to employ GIS. That is for two related but distinct reasons. First, historians, like other humanists, have at times been suspicious of the epistemological origins and uses of GIS. Historians, moreover, have been unconvinced that GIS, a set of methodologies and software packages dependent on precision and certainty, is nuanced enough to represent, much less analyze, the ambiguous and multivalent evidence historians encounter—or that historical data is consistent enough for the demands of GIS. Historians and other humanists insist that tools employed for examining and analyzing historical spaces must match the fluidity and ambiguity of their subjects.

Little of the spatial turn in history has been based on GIS or on sophisticated geographic representations, therefore, but that shows signs of changing. The simultaneous advances in GIS and other geovisualization technologies, the development of large historical spatial datasets, and the turn to place and space among historians present an intriguing opportunity to produce new kinds of scholarship that combine history and geography. It is now relatively easy to transform proliferating spatial datasets of historical phenomena into maps. But it is not easy to produce maps that either reflect or prompt new historical insights and knowledge. Maps that convey or stimulate historical analysis still require the "judgment, discrimination, critical acumen"—the foundations of scholarship—that Wright deemed "of paramount importance in mapping" 70 years ago (Wright 1942, 541).

Geography, for its part, has prepared the way for an alliance with history. Practitioners of GIS are more sophisticated and self-aware than earlier criticisms suggested and than historians may assume. Critical geographers have provided historians with a rich theoretical literature on which to build. Now it is historians' turn. They need to develop their own vocabulary and traditions for dealing with geographic representation.[7]

The interpretation of maps will require a particular kind of humanistic language even as historians learn to work with products of scientific and social scientific methods, a language that speaks with precision of empirical evidence but recognizes that all evidence carries multiple meanings. Historians will need to explicate maps as

[7] See also Alexander von Lünen's paper in this book on the question of historian's genuine use of GIS and Alexi Baker's paper on a more eclectic use of GIS by a historian.

other humanistic scholars read poems, paintings, or films. Maps will need to become agents of discovery rather than mere illustrations. Maps will have to be interpreted through language, in the forms of explication native to the writing of history.

Even as historians build on the humanistic tradition of close reading, they will face a challenge rarely confronted by humanities scholars: interpreting objects they themselves call into being. The texts will not be works of literature or art created by other people but texts that historians themselves help create. The maps will be the product of a question, a frame, and a body of evidence; issues of design will be intrinsic to their form and content. Historians will need to collaborate with geographers and cartographers in a cooperative spirit often alien to the discipline.

Again, geographers have led the way in the creation of bridging language, finding interpretation where it was not always apparent before. Geographer J. B. Harley points out that "our maps are always an *argument*" because "in mapping historical sources we [...] seek to persuade." Historical maps "force a particular reading of reality by suppressing other aspects of the past" (Harley 1989, 85–86). Denis Wood argues that the arrangement of maps in an atlas inevitably tells a story to some interpretive end: "the whole comprises a narrative sequence that is not just one thing after another, but a single thing, an argument." Wood calls on geographers to acknowledge and fully seize the opportunities afforded by the rhetorical, textual, and argumentative nature of map and atlases: "Maps can convey incredible quantities of information with an efficiency and effectiveness that cannot be matched. All that has been lacking has been the will and imagination to exploit this matchless potential in the acknowledged service of an argument" (Wood 1987, 34, 36).

Historians, like geographers, recognize that multiple structures interact with each other in unique ways at any given moment in time in an intricate and ever-changing fashion. One of the present authors has called this patterned yet unpredictable change "deep contingency": "All social life is 'contingent'," Edward Ayers argues, "implicated and unpredictable, because all parts of life depend on one another. What we think of as public and private, economic and political, religious and secular, and military and civilian are deeply connected. Social change can start anywhere and lead anywhere." Deep contingency focuses on the multiple consequences of unexpected actions, with events rippling or surging across space and time. Accordingly, deep contingency can be—perhaps must be—mapped.[8]

Consider, for example, the complex historical process of the end of slavery in the United States during and after the American Civil War. Emancipation did not come in the full historical light, like elections. It did not come on a few days and on a fixed geographic stage, like battles. Instead, it came around the edges of the story. It started before the war began and ended long after the war had come to a close. It happened on dark roads and in obliquely worded government documents. It started and then stopped. It raced ahead and then was retracted. Military events helped it and hindered it. White northerners supported it and resisted it. It was entangled with war making from the very beginning, the motives behind it appearing choked and

[8] Ayers (2010), Ayers (2003), and Ayers (2005, 135).

opportunistic even when they might have been idealistic at their core. Some slaves were able to escape to Union lines within months of the beginning of the war, but vast numbers remained as firmly bound by slavery in 1865 as they had been in 1861; some did not see freedom until after the war had ended, and the Freedmen's Bureau traveled from one plantation to another. Emancipation unfolded unevenly across an expanse the size of Continental Europe, on isolated plantations and farms beyond the sight or reach of outside authorities.

Deep contingency in emancipation flowed up and down through the command structure and social order, back and forth from battlefield to home front, and to and from military and political life, in words and deeds. Emancipation steadily gained momentum, but it repeatedly circled back on itself and refused to run in fixed channels. Emancipation may be one of the deepest and most contingent events in US history—affecting not only the most people in the most profound ways but also through the broadest range of social institutions, including politics, economics, the military, and religion. It was also at great risk, for a long time, of not happening or of not coming to fruition. Its origins and momentum depended on events at least apparently about something else, such as preserving the Union, military victory, and political survival.

We have been working to create ways to represent the deep contingency of emancipation, building on the strengths of GIS software to pull apart complex processes into analytically distinct layers. Emancipation was simultaneously a legal, demographic, and military revolution. Telling the story of emancipation means paying attention to these layers, relating how they warped, intersected, and diverged at crucial junctures. Telling such a story in words would take volumes, but it is possible to visualize this change by animating large, layered datasets at multiple scales. Given the inflexible limits of the printed page, we sometimes freeze emancipation at a single frame, viewing its components individually in order to see the connections between them. Such a process would not be unlike axonometric techniques used by architects and engineers, exploding components apart to imagine complex systems piece-by-piece (Fig. 13.3).[9]

Emancipation was a legal process, through which slavery in different areas and through different provisions was declared impermissible by the United States government. By 1863, the Emancipation Proclamation had removed legal protections for slavery in most of the seceded states. Slavery remained legal, however, in much of the upper South, including Missouri, Kentucky, and Maryland. Here, the

[9]Legal boundaries are drawn from the text of the Emancipation Proclamation and texts gathered in Ira Berlin et al. (1982–1993); data on the distribution of the enslaved population are courtesy of Minnesota Population Center, *National Historical Geographic Information System: Version 2.0*. Minneapolis, MN: University of Minnesota 2011; the heatmap of emancipation events is derived from a kernel density analysis of data on emancipation obtained from reports detailing union military movements, May through September, 1864, as compiled in the United States War Dept. (1880–1901). An animated, interactive version of this map will appear in 2012 (http://dsl.richmond.edu/emancipation/). Thanks to Leslie Rowland for comments on previous versions of this map, which appeared in Ayers and Nesbit (2011).

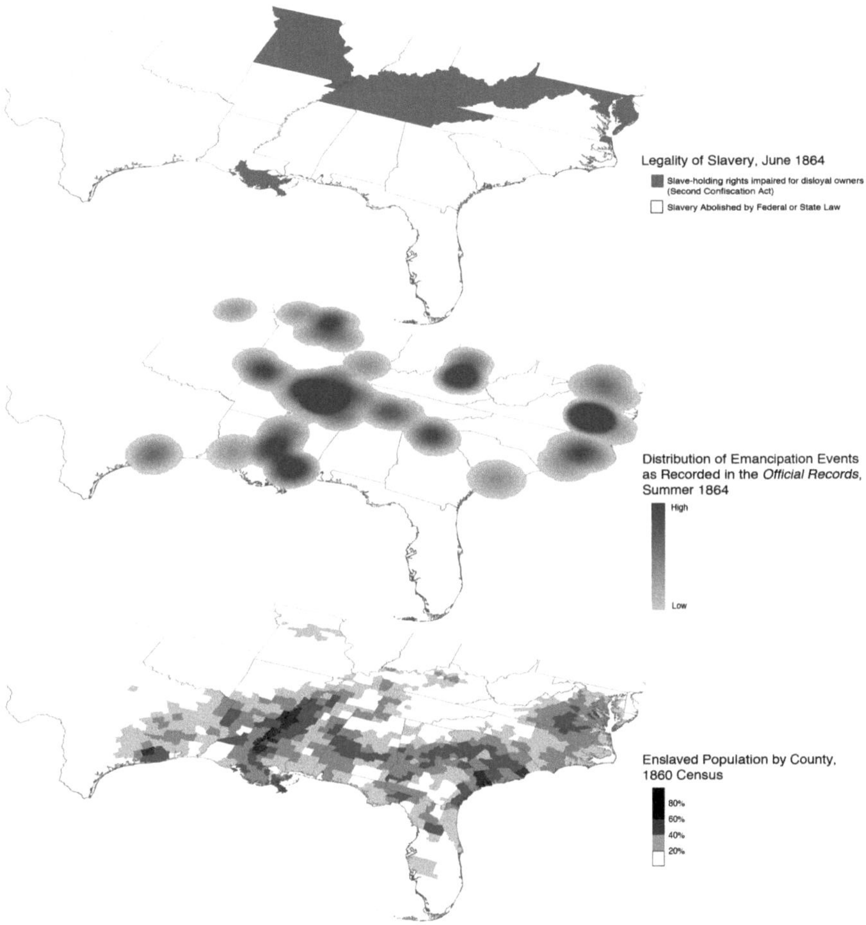

Fig. 13.3 Thematic maps showing data related to slavery in the 1860s in the USA (Taken from Ayers and Nesbit (2011))

institution still met constraints dependent on the loyalty of slaveholders. Through the Second Confiscation Act, slaves of any disloyal owner, regardless of their residence, had claim to freedom.

Emancipation was also a demographic upheaval. African Americans participated in tremendous migrations that cut in multiple directions. Slave owners forced men, women, and children away from the threat of Union troops, far from coasts and major rivers into the southern interior. At the same time, Confederate authorities conscripted enslaved men to work on fortifications, drawing many into battle zones. These conflicting demands pitted Confederate authorities against owners of chattel for the power to determine the movement of black laborers. The black men, women, and children caught up in these movements took whatever advantages they could from these intra-Confederate forced migrations but more often than not suffered hardship from the chaos. Maps reflecting the enslaved population of the United States in 1860, while used by President Lincoln to trace the progress of emancipation and reflected in the map presented here, had expired as anything but the most general approximation of the slave population of 1864.

A third phenomenon touched the migrations and shifting legal boundaries at every point, influencing the decisions of the United States and Confederate governments as well as those of slave-owning refugees. Enslaved men and women, individually and in large groups, continually escaped the direct control of civilian slaveholders and forced servitude to the Confederate government by entering Union lines. Mapping these escapes, as we have by drawing on officers' correspondence gathered in the *Official Records of the War of the Rebellion*, reveals a geographically and temporally uneven process. In the summer of 1864, men and women made such escapes in many parts of the South but especially in Virginia's Piedmont and Valley, along banks of the Mississippi north of Vicksburg, and in the recruiting fields of Kentucky.

No single characteristic, other than the presence of Union troops, marked all the places in which men and women escaped slavery in the Confederacy. The size and stability of nearby Union forces mattered for men and women who sought to become free, and African Americans seemed to have come to freedom more often in newly occupied territory. Yet some places that meet these characteristics reveal relatively few incidents of emancipation. Despite the tens of thousands of US troops preparing for the siege of Atlanta under General William T. Sherman, for example, very few African Americans found freedom there. Exasperated that he could not find enough black men to conscript as laborers, Sherman complained to a colleague, "negroes are as scarce in North Georgia as in Ohio." As he explained, "their owners have driven them to Southwest Georgia." Sherman's lament reflects the contingency of complex geographic and historical processes, how unpredictable patterns emerged from a cascade of decisions. Emancipation emerged at the intersection of escapes, arms, migrations, and law.[10]

As this one example suggests, creating a new atlas of American history is filled with exciting possibilities and productive challenges. Some of the challenges are disciplinary, as GIS adapts to representing change and as history learns of the possibilities and demands of GIS. Some of the challenges are institutional, as departments, institutions, and organizations learn to cooperate in new ways. Some of the challenges are technical, as tools for production and distribution continue to evolve at a dizzying pace.

Just as scholars at the beginning of the twentieth century recognized the great possibilities afforded by new sources of information and new professional capacity, so might we. History and geography have lived through generations of experimentation, self-critique, and growing sophistication over the 100 years since

[10]William T. Sherman to Lorenzo Thomas, June 21, 1864 in United States War Dept. (1880–1901, ser. 1, vol. 39, 2: 132).

John Franklin Jameson, Charles Paullin, and John K. Wright embarked on their great work. Perhaps it is time to embrace the possibilities of our own moment on the historical landscape.

References

Ayers EL (2003) In the presence of mine enemies: civil war in the heart of America. W. W. Norton, New York

Ayers EL (2005) What caused the civil war? Reflections on the south and southern history. W.W. Norton, New York

Ayers EL (2010) Turning toward space, place, and time. In: Bodenhamer DJ, Corrigan J, Harris TM (eds) The spatial humanities: GIS and the future of humanities scholarship. Indiana University Press, Bloomington, 1–13

Ayers EL, Nesbit S (2011) Seeing emancipation: scale and freedom in the American south. J Civ War Era 1(1):3–22

Beard C, Beard M (1927) Rise of American civilization, vol 1. MacMillan, New York

Black J (1997) Maps and history: constructing images of the past. Yale University Press, New Haven

Cappon LJ (1979) The historical map in American atlases. Ann Assoc Am Geogr 69(4):622–634

Cappon LJ, Petchenik BB, Long JH (1976) Atlas of early American history: the revolutionary era 1760–1790. Princeton University Press, Princeton

Carnes MC, Garraty JA, Williams P (1996) Mapping America's past: a historical atlas. Henry Holt, New York

Channing E, Hart AB (1896) Guide to the study of American history. Ginn, Boston

Cook KS (2002) The historical role of photomechanical techniques in map production. Cartogr Geogr Inf Sci 29:137–154

Dobson JE (1983) Automated geography. Prof Geogr 35:135–143

Donnan E, Stock LF (eds) (1956) An historian's world: selections from the correspondence of John Franklin Jameson. American Philosophical Society, Philadelphia

Goodchild MF (1991) Just the facts. Polit Geogr Q 10:335–337

Harley J (1989) Historical geography and the cartographic illusion. J Hist Geogr 15:85–86

Harrower M (2004) A look at the history and future of animated maps. Cartographica 39:33–42

Higham J (1965) History: professional scholarship in America. Johns Hopkins University Press, Baltimore

Ira Berlin et al (eds) (1982–1993) Freedom: a documentary history of emancipation, 1861–1867, 8 vols. Cambridge University Press, New York

Karpinski LC (1934) Review of 'atlas of the historical geographical geography of the united states' by Charles O. Paullin. Isis 22:308

Meinig DW (1986–2004) The shaping of America: a geographical perspective on 500 years of history, 4 vols. Yale University Press, New Haven

Merk F (1933) Review of 'atlas of the historical geographical geography of the united states' by Charles O. Paullin. New Engl Q 6:621

Paullin CO (1932) Atlas of the historical geography of the united states. Carnegie Institution of Washington and the American Geographical Society of New York, n.p.

Schulten S (2011) Thematic cartography and the study of American history. In: Daniels S, DeLyser D, Entrikin JN, Richardson D (eds) Envisioning landscapes, making worlds: geography and history. Routledge, New York, pp 55–61

Tanner H, Reiff J, Long JH, Hoerder D, Dobyns HF (eds) (1995) The settling of North America: the atlas of the great migrations into North America from the ice age to the present. Macmillan Press, New York

Taylor PJ (1990) Editorial comment GKS. Polit Geogr Q 9:211–212
Thomas WG, Ayers EL (2003) The difference slavery made: a close analysis of two American communities. http://www2.vcdh.virginia.edu/AHR/. Accessed 02 Dec 2011
Turner FJ (1925) Significance of the section in American history. Wis Mag Hist 8(3):255–280
United States War Dept (1880–1901) The war of the rebellion: a compilation of the official records of the union and confederate armies. Government Printing Office, Washington DC, published in 4 series (70 volumes in total)
Wilgus AC (1932) Review of 'atlas of the historical geographical geography of the united states' by Charles O. Paullin. Hisp Am Hist Rev 13:232
Wood D (1987) Pleasure in the idea/The atlas as narrative form. Cartographica 24(1):24–66
Wright JK (1932) Sections and national growth: an atlas of the historical geography of the united states. Geogr Rev 22(3):353–360
Wright JK (1942) Map makers are human: comments on the subjective in maps. Geogr Rev 32(4):527–544
Wright JK (1947) Terrae incognitae: the place of the imagination in geography. Ann Assoc Am Geogr 37(1):1–15
Wright DJ, Goodchild MF, Proctor JD (1997) GIS: tool or science? Demystifying the persistent ambiguity of GIS as 'tool' versus 'science'. Ann Assoc Am Geogr 87:346–362

Chapter 14
Tracking in a New Territory: Re-imaging GIS for History

Alexander von Lünen

The story is entirely true, because I imagined it from one end to the other.

(Boris Vian: L'ecume de jour, 1947)

14.1 Introduction

There is much talk lately about geographically referenced, digitised historical information. Websites such as *Historypin* or *Vision of Britain* draw much public and popular interest, while "space" has become the buzzword for many historical studies in recent years. As with all the preceding disciplinary "turns", "spatial turn" historians are looking beyond their discipline's boundaries to find inspiration and impetus for their work. However, like with those other turns, it remains to be seen whether spatial history and geographical information systems (GIS) will have a lasting effect on the academic subject of history. While all these turns (iconic turn, linguistic turn, cultural turn, etc.) have made some impression on history, they have asserted its main methods rather than contesting it. And why would they? Probably no other academic discipline has been challenged by "outsiders" like history. Historians have been keen to listen to these proponents from other fields and joining them on an adventurous trip just to return home unscathed, just as Dorothy returned home from Oz, realising: "The next time I go looking for my heart's desire, I won't look any further than my own backyard; if it's not there, then I never really lost it to begin with".[1]

[1] *The Wizard of Oz* (1939) (PAL DVD, Warner Home Video, 2006) ASIN: B00005NMWB, run time: 101 min.

A. von Lünen (✉)
Geography Department, University of Portsmouth, Portsmouth, UK
e-mail: alex.von-luenen@port.ac.uk

A. von Lünen and C. Travis (eds.), *History and GIS: Epistemologies, Considerations and Reflections*, DOI 10.1007/978-94-007-5009-8_14,

It is therefore amusing to read and hear how GIS is supposed to radically change the way history is going to be done hence. Ironically, historians seem to have forgotten that computers and geography had been introduced to historical studies over 40 years ago.[2] That we haven't managed to get past the stage of some "tentative beginnings" (Bullough 1966/1967) but are still discussing how we get "mainstream" historians to use computers for more than just word processing speaks volumes. That most historians do not feel inclined to go out and venture into GIS may be due to their lack of technical prowess, as many historical GIS activists frequently lament. But it may also have to do with the fact that the basic methodology has not changed much since computers were first used by historians in the 1960s. To be fair, this is being addressed within the Historical GIS community, and alternative analysis and visualisation methods are being sought. Yet, one has to wonder whether "smarter" software is indeed the only way out of this cul-de-sac.

When Dorothy encountered the scarecrow on the way to the wizard, she was puzzled by his obsession to obtain a brain. "What would you do with a brain if you had one?", she asked. To which the scarecrow replied: "I would become a doctor of thinkology!" If the historical GIS community simply keeps striving for more "brains", i.e. more sophisticated software, I am afraid they won't make much headway in many history departments. In media theory, there is the golden rule that every new medium has to imitate established media before it can evolve into its own sphere. Photography, when it appeared in the nineteenth century, was first and foremost used to portrait and capture "reality", imitating painting and engravings—who actually were not depicting the real thing but rather the artist's perception of their subject—thereby freeing painters from the arduous task of mimicry, giving them more freedom to leave realism behind. Likewise, GIS for some time imitated traditional cartographic techniques and is still adhering to its principles while having added a number of analysis and visualisation techniques to its toolbox.

I think historical GIS should have passed through the "imitation stage" by now, i.e. simply producing maps through quantitative methods and data should not satisfy anyone anymore. On the other hand, developing new software while the potential of the old one has not been explored (let alone exploited) yet seems nonsensical at best and will degrade historical GIS into a mere "Look Ma, no hands!" stunt. It is about time that we discuss how to relate GIS to actual historical scholarship, rather than lamenting about the lack of proper software or the lack of proper historians.

14.1.1 Definitions. Or: Is GIS Just a Four-Letter Word?

Before the main points are outlined in this chapter, it would be a good idea to define some of the terms that will be discussed. The word "GIS" evokes a number of different concepts for different people, and this fact alone justifies a paper of its own. In light of Wittgenstein's verdict that the meaning of a word is determined through

[2]Cf. my interview with Le Roy Ladurie in this book.

the use of language (Wittgenstein 1978, 344), however, most people use "GIS" as a synonym for a software product produced by a company from Redlands, California. Looking up the definition in a dictionary, though, reveals a much broader concept than is usually contemplated. The *Oxford Dictionary on Computing*, for example, defines GIS as "an information system for providing users with information about objects and features in their geographical contexts" (Daintith and Wright 2008, 219). It is thus clear that GIS can refer to a very broad range of data and systems and that the popularity of certain GIS products does not overshadow the fact that these products are just one manifestation of a GIS.

Furthermore, I would like to go one step further than Jack Owens, who discourages the use of the term "Historical GIS" for historical studies using GIS. "Historical GIS" (as a concept) seldom appeals to historians, not only because they follow the epistemology set out in historical geography, but also because they are usually more about creating data repositories of historical-geographical data, i.e. HGIS are often—if not most of the time—about what archivists do, rather than what historians do. Archives are important to historians, obviously, but they are normally the users rather than the creators of one.

In terms of epistemology, a specifically tailored HGIS is not any different from using a standard desktop GIS, however. HGIS usually are distinct from regular GIS in their approach to storing historical data, rather than facilitating a novel way to analyse data. Therefore, as far as the user—that is, the historian—is concerned, HGIS is not much different from any other GIS. I will consequently draw no dividing line between them and will simply to refer to all such software as "GIS"; whether the software in question was specifically designed to deal with historical data is irrelevant from the historian's perspective. In the course of this chapter, I will explain why.

It would be in order, however, to differentiate between GIS projects that are purely created for storing historical-geographical data and those that use GIS to facilitate historical research. My chapter is not so much concerned with the former, the archival form of GIS, but rather with the latter: the "research" GIS. For the sake of my argument, I will use the term "Historical GIS" (HGIS) for said data repositories and use the term "Gistory" to refer to "GIS in historical research".[3]

14.2 Historiography and GIS

When we speak of GIS in history, and that Gistory should take some lessons from historiography, what history are we actually talking about? A quick query for all courses that have "history" in their title on the website of UCAS, the British

[3]This neologism—gistory—doesn't seem to have been used much yet, as far as I could make out. The only one using it seems to be a German webportal dedicated to historical geography and GIS: http://www.gistory.de/, accessed 20 Sep 2011.

governmental organisation that is responsible for managing applications for higher education courses in the UK, resulted in 57 different courses.[4] Many of these historic sub-disciplines have developed their own canon of methods, and it would be a good idea to have a quick review of the general methods that are found across these sub-disciplines, and how they relate—or can be related—to Gistory.

14.2.1 Empiricism and Positivism

The study and practice of history has experienced many challenges and changes in its evolution. While history was regarded as a form of literature in premodern times, the scientification of scholarly subjects in the nineteenth century did not exempt it. Philosophers such as Kant and Hegel had (and still have) exerted a great degree of influence on the formation and self-image of the subject, but it was probably German historian Leopold von Ranke (1795–1886) who raised the stakes for the subject. His verdict that history should show the past "how it really was" has often been misinterpreted. His basic point of departure was to abandon the literary form of history that had prevailed thus far. History had traditionally embraced mythology and folk tales, and Ranke demanded that these tales as source for historical knowledge be left behind and that "primary" sources should be chiefly employed, i.e. documents from archives.

Ranke is regarded contemporarily as objectivist history's posterboy, implying that he was a stalwart reconstructionist who believed that the "truth" about the past can be arrived at with an almost scientific credibility if one just sticks to the sources and separates fact from fiction, something that was disputed—if not mocked—later.[5] Hungarian philosopher György Lukács (1885–1971), for instance, stated that Ranke's historiography would present nothing but "a collection of exotic anecdotes".[6] This is unfair, though. Ranke's aim was first and foremost to establish

[4] Some of these were a bit redundant (e.g. "art history", "art history studies" and "history of art") and reflect a non-uniform naming scheme across UK universities. The query on http://www.ucas.com/ was done 9 May 2008, when the idea for this paper was born.

[5] The terms "objective history" or "reconstructionist history" are not really interchangeable, yet are quite often used in concert. "Reconstructionist" history tries to meticulously reconstruct an account of past events by using as many sources as possible, but this doesn't necessarily reflect that the account thus produced actually represents the "truth" about the past. "Objective" history, on the other hand, doesn't have to be reconstructionist. Rather, "objective history" tries to establish as many "facts" as possible, but not necessarily by way of reconstruction (although this usually plays a great role in objective history). Quantitative history, for example, often regards itself as "objective", for its methods and findings are "scientific" and retraceable. Both the "objective" and the "reconstructionist" school of history are not per se synonymous with "positivistic history", but in practice all three terms can be regarded as different instances of the same methodological system. For more on this, see Iggers (1997) and Munslow (1997).

[6] From R.W. Davies's "Notes toward a Second Edition" of Carr's "What is History?"; (Carr 1990, 159). I have noted that there is a profound *lost-in-translation* issue in many English-language

(for his days) a critical method of historical scholarship that would critically examine sources and contextualise them. While this approach is open to criticism, it must be noted that it instigated a good deal of progress in the subject of history and remains to this day one of the building blocks of the discipline.

What most critics actually focus on is the often encountered positivism in the so-called "objective history". "Positivistic history" believes that "historical evidence can be discovered, evaluated and objectively constituted as facts". Positivistic history strives for "discoverable and repeatable laws" by which human society can be categorised and analysed (Munslow 2000, 187).

Historical studies using GIS, unfortunately, are still imbued with this notion of scientific objectivity. Be it that the disciplines of social and economic history—the historical sub-disciplines that seem to be most enticed by GIS—see positivism as their natural choice, be it that standard GIS packages usually only offer statistical analyses and its visualisation as choropleth maps which demands categorisation, it should be obvious that results from a GIS analysis/visualisation must be placed into the proper historic context, commencing with being more critical about the sources (a.k.a. data) that are put into the GIS.

14.2.2 Postmodernism

On the other end of the methodological spectrum—opposite positivism—is "post-modern history", a subject of even greater controversy. Where empiricist history acknowledges the subjectivity of historic narratives and that it may actually never be possible to arrive at the "objective" side of things, postmodernism goes a step further and denies that the principle possibility of knowledge about the past can be retrieved from the sources. Whereas Leopold von Ranke attempted to depart from the use of myths and folk tales in history, postmodernists such as Hayden White or Roland Barthes argued that history is no different from poetry, i.e. that there is no such thing as a separation of fact and fiction (Iggers 1997, 100). Historical narratives, they argued, are not a statement on the past but are always the construction of it. History is therefore "itself a form of the 'fictions' and 'myths'" empiricists tried to abandon, according to White and others (Burke 2009 [2005], 115).

This approach to history obviously drew criticism, not least because post-modernists typically came from outside the field of history and were therefore perceived to be too unfamiliar and insensitive to historiography and also because of

publications. Stephen Davies, for example, translates Ranke's "Wie es eigentlich gewesen ist" as "to show what actually occurred" (Davies 2003, 28). This is inaccurate, though. The German word "gewesen" would in most cases translate into the English word "was", whereas the German equivalent of "occurred" would be "geschehen". However, the whole sentence could also be translated as "what the essence of it was", and considering Ranke's time and work, this would be a much more accurate translation. This is a point Bentley makes and which I—as a native German speaker—can wholeheartedly subscribe to (Bentley 1999, 39).

postmodermism's political and moral implications. Postmodernism was primarily a philosophical attack on modernist's claim (such as from Francis Bacon, 1561–1626) that we may arrive at an ultimate truth if we just try hard enough or that there is at least an ultimate truth out there, yet humans might never be able to grasp it. Postmodernism denies this claim, arguing that there is not *the* truth but many and every one of these truths is valid and solely rests upon cultural codes. This "relativism" fundamentally questioned the validity of our knowledge, for we could never gain universal knowledge but just an interpretation of our observations.

More conservative historians, such as Keith Windschuttle, have not only attacked postmodernism as redundant at best—as no serious scholar would take his or her theses as universal knowledge—or morally despicable at worst.[7] As Georg Iggers reports, the relativism that is implied in postmodernism leads to extreme moral difficulties in history—and to contradictions in postmodernism. One accusation made by Windshuttle and others is that the historical relativism would vindicate abhorrent political propaganda such as Holocaust denial—in the vein that, if there is no truth in history, why should this particular history (the Holocaust) be true? Iggers points out that Hayden White got caught up in some torturous explanations, when he admitted that it is morally wrong "to deny the reality of the Holocaust", but "impossible in a historical narrative to establish objectively that it happened" (Iggers 1997, 13).

More recent accounts on historical scholarship strike a somewhat more moderate tune. While it is accepted that postmodernism made some important points about the nature of knowledge of the past, its denial of *any* knowledge and history as pure fiction has been rejected. Most historians try to find a middle ground, redefining the "classic" hermeneutics of historians such as Ranke or Dilthey while not abandoning empirical principles and maintaining that despite all legitimate criticism made by postmodernism, the practice of historical inquiry has established a canon of "inter-subjectively verifiable methods".[8]

14.2.3 Which Methodology to Use with GIS?

Employing GIS in historical studies is thus not a matter of "which side are you on?", but rather of using GIS to critically engage with one's sources and inferences, i.e. to introduce into Gistory what has become good historic craftsmanship.

Eventually, all these "turns" we hear about so much in the humanities, such as the "cultural turn" or "spatial turn", in the end present themselves as some kind of rebranding effort. Leaving aside all these debates on positivism, postmodernism or objectivity, the basic methodology of the historian remains unchanged: going to the archive, trawling through old documents, comparing and distilling them and writing up one's conclusions. Meaning: Whether empiricism is cherished and endorsed, it

---

[7] See Windschuttle (1997) for a refutation of postmodern history.

[8] Middell (2008, 106); see also Burke (2009 [2005], 127). See Ginzburg (1999) for a general attempt to find a middle ground between positivistic and postmodern history.

is still the bread and butter of the trade. Not even the most hardcore postmodernist historian would go completely without source-based work. Or as Davies puts it: "[O]n closer examination it turns out that in most cases these new departures are actually the application of empirical methods to new areas, and they reveal its extension rather than its demise" (Davies 2003, 42).

One of the central concepts in historical scholarship is that of the "historical narrative". No matter which school you are with, dealing with a narrative is inevitable; the main difference being *how* you deal with it rather than *that* you are dealing with it. There is yet some disagreement as to what comprises a narrative or rather what constitutes it. For some scholars, "a narrative requires a narrator (i.e. the person telling the story)", but for others, "the narrator is one of a number of instruments [...] that can be used in the narrative process of representing events" (Abbot 2002, 13).

I have argued previously that GIS can act as a historical narrative (cf. Lünen 2010a). I go beyond what Staley referred to as "visualization as an alternative to prose" (Staley 2003, 35ff) in that I argue that visualisation is just one way of creating a narrative but that the analytic capabilities offered by a GIS constitute a narrative as well, "tak[ing] narrative to mean all modes of conveying stories" (Abbot 2002, 16). And this conveyance of stories doesn't necessarily mean conveying the story to an audience but rather the exploration of the narrative by the "gistorian". When postmodern historians argue that there are infinite numbers of narratives out there, that it is therefore impossible to find the only true and valid one and that narratives are thus constructed (i.e. arbitrary), we can construct narratives with a GIS and explore their plausibility.

Before I expand on this further, it would be appropriate to examine the differences between geography and history and how the two disciplines handle narratives.

14.3 History and Geography and Where It Leaves History and GIS

It has become a somewhat fashionable trend to stress the relationship between history and (human) geography under the umbrella term "spatial turn" in recent years. Yet, as Doering rightfully remarks, the "spatial turn historians" have an almost romantic and nostalgic understanding of human geography, often referring to traditional concepts of the field, rather than the state-of-the-art (Döring and Thielmann 2008a, 19). This is, as Piltz outlines, to do with the vague definition of "space" by spatial turn historians, something eminent German historian Reinhard Koselle k (1923–2006) criticised as early as 1986 (Piltz 2008, 78). "Space", it seems, is used metaphorically in many historical circles rather than actually delineating a scholarly concept and application.

When historians employ spatial concepts, these are often borrowed from human geography, namely "imaginative geographies" and "spaces of identity, which are often articulated through spatial images such as mobility, location, borderlands, exile and home" (Blunt 2003, 75–76). It is obvious that such imagined spaces are quite hard to map in a GIS—unless one strays from the trodden path that a certain Californian company has left for us to follow, that is. Alternative approaches have been contemplated, not least by this author.[9] Historic accounts rarely have mathematical coordinates in them, rendering a direct geographic analysis futile. Rather, most sources have relative geographical descriptions in them, i.e. a positional description relative to another location, such as "50 mi north-east of London".[10]

"Georeferencing!" some of you might now cry out (cf. Hill 2006). Putting aside the tediousness and costliness of such an enterprise (considering that there may be hundreds or even thousands of places that have to be looked up in address registers, maps or cadastres), are the *places* we deal with in history always physical and of a material quality? Certainly not. How many sources may contain references to the "Garden of Eden" in it—a place, but not a *real* place—or, even more plausible, how many historical documents have a placename in them, yet do not refer to the geographical location as such, but rather to its cultural connotation? Just think of "Jerusalem" to imagine how many ways a placename creates some kind of response, depending on your own preferred political or religious world-view; and more than often also depending on the historical context, such as the eleventh century crusades or the twentieth century Middle Eastern conflict, in the case of "Jerusalem". Geographer Yi-Fu Tuan notes: "distinguish[ing] between places that yield their meaning to the eye, and places that are known only after prolonged experience". The former he refers to as "public symbols", the latter as "fields of care" (Tuan 1979, 412).

Geography—or cartography, to be more precise—on the other hand has for a long time indulged in a positivism that is staggering; cartography has stylised itself into a "mathematical-geographical, or purely scientific subject" (Schlögel 2009, 90), oozing a scent of objectivity since abandoned in other fields of humanistic inquiry. Unfortunately, many works in the field of Historical GIS are still indulging in positivism. Examining papers in the SSHA's journal *Social Science History* that use GIS (and there are many such papers), the majority are rather descriptive, using GIS to carry out geo-statistical analyses and to produce thematic maps. The critique of Marble and others that many GIS applications are simply used as a map making tool, or to "creat[e] new maps from old" with it, has not lost much of its actuality and legitimacy in this respect (Marble 1990, 13).

Make no mistake: Maps are essential sources of historical knowledge, just like textual documents; they are the "accumulated knowledge about the spatial form of the world in which we live" (Marble 1990, 9). Many of the cartographers, geographers and Historical GIS users have yet to engage in source criticism, as much as historians have yet to arrive at a "topographical hermeneutics" that German historian and writer Nicolaus Sombart (1923–2008) wrote of (Sombart 1992). He criticised that cultural studies (including history) focuses too much on

[9]See Aucott et al. (2009) or forthcoming publications on the Leverhulme Trust-funded project mentioned earlier in the book.

[10]In the GBHGIS, this has been tackled by using a data model that allows for relating places by such relative information, rather than by coordinates. See Southall et al. (2009). See also Baker's chapter in this book.

the demystification and deconstruction of historical narratives, complaining that this postmodern hermeneutic is "without place" and that the "sensual concreteness of life" is "not taken seriously" (Sombart 1992, 107). The methods and ideas are all available to contemplate a "topographical hermeneutic", but Sombart attacks the "methodological pluralism" of the "interdisciplinary humanities" as the reason why this has not occurred yet (Sombart 1992, 108). For Sombart, everything has a place, and history can't just be discussed in a "spiritual realm" that is "detached from the sensual world".[11]

Taking such a position he was—and still is in many cases—ahead of most of his fellow historians; in geography this was also contested at the time.[12] It is exactly what Sombart asks for in his call for the "topographical hermeneutic": to accept that some places are bigger than others or that "places are conceived of as being unique rather than singular" (Castree 2003, 182). Sombart used the example of the small Swiss city of Ascona as an example to discuss the uniqueness of places. While marginal in size (around 5,000 inhabitants nowadays but only half of that in mid-twentieth century), the town not only has been an extremely popular tourist attraction for decades, it also became a magnet for artists and meeting point for intellectuals during the twentieth century. Sombart points out how certain towns—or just parts of a town such as the part of Munich named Schwabing—have an image of vibrancy hovering over them that attracts many people. Subsequently, such a (self-) image becomes a self-fulfilling prophecy. He applies the term "psychotope" (a term he borrows from architect Richard Neutra) to it in reference to the term "biotope", stressing the fact that "every place(-name) has a cultural connotation" and an "aura of associations" attached to it (Sombart 1992, 113–114).

How does this discussion help us with using GIS in history? The point of the uniqueness of places is well established with many historians, and the dealing with special places such as Ascona does not hold much interest outside cultural history.[13] Historical GIS following the digital gazetteer approach—such as the Great

[11] Sombart (1992, 107). I wholeheartedly agree to this sentiment of Sombart. From my experience, it is exactly this *sensual* aspect of GIS that makes it so attractive and fruitful for history, especially in teaching and communication. In a GIS course for history students a few years ago, for example, we used the German package "Karten-Explorer" (http://www.bfav.de/kartenexplorer/), a simple software that has only a few features one would expect from a GIS. Yet, the students were quite enthused because it gave them a way of visualising historical data in a map, and they felt they can relate better to that material for they had something familiar (a map) to look at (see the contribution from Mares and Moschek in this book). For the *Vision of Britain* website, to use another example, we found that the largest userbase are local and family historians, who were interested in the place/area they or their ancestor were from. Again, the sensual angle a GIS offered makes a great difference. People are interested in things they can relate to, and the success of the *Vision of Britain* website is due to its relatable content (the website has c. 70,000 unique user every month). I certainly don't dismiss the more abstract approaches to history, as I am no stranger to *theory*, but gistorians are well advised to keep in mind that this more intuitive approach to history may yield more innovative ideas than staying in the confines of a theory.

[12] See the interview with Gunnar Olsson in this volume.

[13] Although "cultural" seems to be the buzzword of the day in historical studies lately.

Britain Historical GIS and its web interface "A Vision of Britain through time"[14]—could easily implement this notion of uniqueness by storing more qualitative place information such as narratives of place. In "Vision of Britain", for example, this is contemplated by the mark-up of historic travel narratives, in which place names within the narrative are linked to the entry for that place in the database, so that the narrative can be compared to data from census reports. This could be extended to various other materials—historic photos, interviews, etc.—creating an archive of disparate place-bound narratives that allow historians to compare them and produce interpretations of historic accounts. Consequently, there are websites who make a lot of use of these disparate sources.

14.4 Epistemology

This discussion leads us to the core of my chapter—or of this book for this matter. What is the insight we gain in historical studies by using GIS? Scholars from the Historical GIS community are often heralding the facilities of GIS to collate and visualise historic data, yet what this yields in terms of gaining new perspectives on historical narratives remains unclear. As mentioned, historical studies conducted with GIS—as opposed to those Historical GIS that primarily deal with creating a data archive—usually make little effort to undertake analyses outside the quantitative approach. While quantitative methods are surely a valuable tool to gain insight into historical processes, they too often are not part of a bigger methodological framework, i.e. many papers that employ GIS in historical studies are surprisingly uninformed about the subject they are dealing with and provide very little inferences beyond mere descriptions of the data being analysed.

For fields such as economic history, this might be understandable given "that [it] likes to pride itself on hard reasoning and exact proof" (Landes 1994, 637). There might be nothing wrong with this sentiment per se, but it is remarkable that historical studies using GIS in this field tend to ignore significant debates in historiography, such as the micro-history school that emerged in the 1970s. Carlo Ginzburg, in the introduction of his famous monograph *The Cheese and the Worms* quite explicitly attacks the notion that social history can only be done by quantitative methods (Ginzburg 1976, xx). Micro-history nowadays is an established technique in historical research, but gistory seems to remain unaffected by it.[15] Micro-history

[14]See http://www.visionofbritain.org.uk/. It should be noted that I worked for this project for a couple of years. See also our paper on the topological data model that was designed for the GBHGIS to allow a potentially infinite number of data sets to places: Aucott et al. (2009).

[15]The discussion of micro-history is beyond the scope of this chapter. The reader is referred to Burke (2009 [2005]) or Levi (1991) to learn more about it. At the European Social Science History Conference (ESSHC) 2010 in Ghent, Belgium, I have called for a "Micro-Historical GIS" in my presentation there. This is but yet another facet of my ideas of the epistemology of GIS in history, so I won't repeat this concept here and rather deal with the general idea. For quantitative history

works from the premise that the (social) history of common individuals can be inferred despite a lack of sources for not so prominent persons by applying techniques from anthropology to historical studies in many fields (a point of view human geography has also adapted).[16] While the social and economic historians in the Historical GIS community might be excused through their sub-discipline's historiography, I find it astounding that geographers involved in the field seem to be untouched by one of geography's oldest tradition: that of exploring.

14.4.1 Exploring

Though the days of "Geography as exploration" (Heffernan 2003, 6) may be past, as the last "white spots" on the globe have been filled in, exploring in the form of field trips is still a vital part of the geography curriculum—both in human or physical geography. Geography, despite the revolution of GIS in recent years, is still very much concerned with going out and looking out for things—a habit that historians have lost, if they ever had it, according to Schlögel: "Historians who close their eyes are like architects who live in Victorian houses but on the other hand preach to everyone else to live in council estates" as he poignantly writes (Schlögel 2009, 272). As mentioned above, many historians do not have much of an eye for sources other than in textual form, with historic photographs and maps appreciated more for their illustrative value rather than as epistemes. This makes it all the more worrying that the geographers in the Historical GIS community do not bring their own subject's tradition of exploration into the mix.

When historians go to archives, they explore the sources to discern the significant from the insignificant documents and to get a general overview of the documents available for the topics in question. In geography, many areas of research are now guided by data acquired by way of remote sensing. While techniques such as data mining are now employed by geographers to sift through the enormous amount of

---

vs. micro-history, see also my interview with Emmanuel Le Roy Ladurie in this volume, which outlines that picking one of these two can sometimes be decided by rather mundane reasons.

[16]The word "micro" in this context refers to different levels, however. First, it concerns the social level, i.e. the "(wo)man on the street" rather than celebrities. Second, it also refers to the geographical and temporal scope. Where many historical studies do nationwide histories over a long stretch of time, micro-history is often only about a couple of days in a village. This is first and foremost due to—third—the scope of the sources available. Where national histories use government reports, for example, to draw the bigger picture, micro-history started from the circumstance that one can find very interesting historical documents that are very limited in scope. Ginzburg, for example, in *The Cheese and the Worms* used records from an inquisition trial against a miller from northern Italy to produce an account of Renaissance cosmology. The miller had been tried for his rather weird model of the cosmos. All the information Ginzburg had about the miller were the inquisition records, which did not reveal much about the miller (other than his interrogation), but Ginzburg used this to discuss this trial in the context of wider cosmological debates, among other things, at the time. Micro-history is therefore not supposed to replace "macro"-history but to complement it. See Levi (1991) for more.

data thus obtained (a page historians should take out of geography's book) this side of exploration is not seen much in Historical GIS, although—as I think—it would be an excellent tool for exploratory data analysis in historical research.

Exploration, by the dictionary's definition, is to conduct "a first or preliminary" study, chiefly to "examine minutely [...] esp. for diagnostic purposes" (Gove 1976, 802). In the context of data analysis, "[e]xploration [...] emphasizes development of theory from data" (Stebbins 2001, 5). Or put in a manner much more in line with what Ginzburg said about the process of research (see below): "Exploratory data analysis is detective work" (Tukey 1977, 1). Exploratory data analysis (EDA)[17] is a set of techniques to discover patterns in data that relies heavily on the experience of the researcher.[18] EDA is not about a single best tool, and it is also not about using it as a *black box* approach, where one inserts data and waits for a result to pop out (cf. Tukey 1977, 3). The output of an analysis thus produced is merely a point of departure for further inquiries, "a subject for thinking" (Andrienko and Andrienko 2006, 165). As Chong Ho Yu points out, EDA and abduction (see below for "abductive reasoning") have a lot in common. Both are trying to find patterns, explore hypotheses and discern the less plausible ones (Yu 2011). Just like historians in archives, as described above. While EDA might not—not yet—have the right set of tools to handle historical-geographical data well enough, Gistory could learn a lot from this approach—or at least from its general notion. Exploring is not simply about finding clues, it is also about engaging with them, i.e. critically examining them, drawing upon that information, and looking for further clues, once new clues have been discovered, revisiting the old ones to check that they still fit in, etc. "When something interesting is detected, new, more specific questions appear, which motivate the analyst to look for details. These questions affect what details will be viewed and in what ways".[19] The exploratory devices GIS offers to history are—in my opinion—one of its best and most attractive features.

By that I obviously don't mean virtual reconstructions of historic places to explore them—as some gistorians have done or proposed. The past is gone and there is no way of bringing it back to examine it in situ.

14.5 History Is Exploration

What I mean by "exploration" in this context is the application of the exploratory spirit from geography to historical research, using GIS to gain an understanding of historical sources and narratives and keeping an open mind about hypotheses. That

---

[17]It should be noted that EDA and data mining are often used interchangeably, while they are historically different strands. The purpose of both, to discover patterns in data, is strikingly similar, though, so for the purpose of this paper, I will treat them as equals.

[18]Tukey (1977, 1);Stebbins (2001, 21)

[19] Andrienko and Andrienko (2006, 4). This corresponds to the concept of "explorative learning" in pedagogics. See Moschek and Mares in this book for more details.

is, I refer to a concept of GIS and history that is comparable to what anthropologists refer to as the oldest form of science: tracking. Italian historian Carlo Ginzburg has introduced this idea into historical scholarship in the late 1980s, and I would like to apply the notion of his "clue paradigm" to the usage of GIS in historical studies.

14.5.1 History Is Tracking

Hunters tracking the game gain their inference about the game (and their hunting strategy) from the traces the game has left behind (usually footprints). From that spoor, a good deal of information can be inferred: species (from the shape of the paw/hoof), size—and therefore often the age—of the animal (judging from the size of the footprint) or possibly whether the animal is hurt or not (irregularities in the steps, one footprint not as deep as the others, indicating a hurt leg, etc.), to name but a few. Police detectives work similarly: from the evidence gathered at the crime scene (or the information delivered by the crime scene investigators), he or she will make inferences about what happened, how many persons have been involved, what weapon was used, whether there was a fight, etc. Probably, the most famous fictional investigator making extensive use of inferred information is Sherlock Holmes, but every investigation—be it criminal or scientific—is based on the principle that most clues discovered are "imperceptible to most people" (Ginzburg 1989, 97–98).

Through the different stages of tracking—anticipating, reading, interpreting and classifying spoors—the hunter in the woods or the detective at the crime scene experiences a "retrospective observation", i.e. in a case where the actual occurrence is in the past (game traversing through an area, a crime happening before the police could arrive), the "causes cannot be reproduced, there is nothing to do but to deduce them from their effects" (Ginzburg 1989, 102/117).

Ginzburg applies this principle to the subject of history. History, he argues, has always been an "individualising" science and while other subjects have developed into quantifiable and reproducible disciplines since the seventeenth century, the "scientific revolution" has led history to become aware of its outspoken commitment to the individual (Ginzburg 1989, 106). Ginzburg likens history to medical practice in this respect: physicians use general knowledge about diseases to analyse a patient's actual malaise.[20] "As with the physician, historical knowledge is indirect, presumptive, conjectural" (Ginzburg 1989, 106). Ginzburg calls this the "clue paradigm" of "diachronic sciences" such as history, archaeology or geology (Ginzburg 1989, 117).

Epistemologically, this "clue paradigm" refers to the principle of "abductive reasoning" that American philosopher Charles Sanders Peirce (1839–1914) deliberated. Peirce argued that beside the two established principles of inferences—induction and deduction—there would be a third: abduction.

[20]Ginzburg also draws a parallel between Sherlock Holmes, Sigmund Freud and Giovanni Morelli in terms of medicine. As he pointed out, all three were trained physicians (Holmes's creator Conan Doyle was a physician before becoming a writer). Ginzburg terms their methods "medical semiotics": "diagnoses [...] inaccessible to direct observation" (Ginzburg 1989, 102).

14.5.2 Abduction

Charles Sanders Peirce developed this concept of inferences over many years, giving it various names and eventually arriving at "abductive reasoning" or simply "abduction". He referred to this as "an inference from a body of data to an explaining hypothesis" (Burks 1946, 301). While "induction" describes an inference from a sample to the whole and "deduction" infers from the whole to the individual, "abduction" is about inferring an explanation for an observed process. In science, Peirce explicated, induction is used to test hypotheses, deduction is used to verify them and "abduction includes the method of discovering them" (Burks 1946, 301). Abduction is therefore the "process of forming an explanatory hypothesis" (CSP CP 5.171).[21] Or as Harman puts it, it is an "inference to the best explanation" (Harman 1965, 88). According to Harman, abduction is the process of finding the most likely explanation for a hypothesis, inferring "from the premise that a given hypothesis would provide a 'better' explanation for the evidence than would any other hypothesis, to the conclusion that the given hypothesis is true" (Harman 1965, 89).

Peirce claimed that induction alone could not generate meaningful new insights, because mere observations would not lead to the "bottom of things": "[Abduction] is the only kind of reasoning which supplies new ideas, the only kind which is, in this sense, synthetic" (CSP CP 2.777).

Deduction and induction are for confirming ideas. They operate on the level of probability, either through quantitative or qualitative methodology, whereas abduction "is based on the principle of theoretical plausibility" (Wirth 1995, 411). Peirce maintained that the evolution of modern science "was in the first instance mere guess-work, or at least mere conjecture" which "was derived from experience". So, according to Peirce, to come up with a new hypothesis, one has to build on prior experience. Induction is then employed to find out "what proportion of its anticipations will be verified" (CSP CP 2.755). Abduction is thus an "original argument", a "prediction in regard to a certain horizon of expectations" (Wirth 1995, 405). It is, according to Peirce, for abduction "that we have any knowledge whatsoever", as all other forms of inferences do not create knowledge (Burks 1946, 302–303).

Peirce's concepts are based on signs: "We think only in signs", as he maintained (CSP CP 2.302). He is consequently regarded as one of the founders of semiotics, the study of signs. Peirce, other than Swiss linguist Ferdinand de Saussure (1857–1913), defined a typology of signs, or "'modes of relationships' between sign

[21] It has become common to refer to the *Collected Papers* (CP) of Charles Sander Peirce (CSP) by the volume number and the number of the paragraph, i.e. a reference such as "CSP CP 2.229" refers to volume 2, paragraph 229 of the Collected Papers of Charles Sanders Peirce. For the Collected Papers used in this paper, see Peirce (1960a) and Peirce (1960b). It should be noted that Peirce himself wrote a paper about the "logic" of historical research, *The Logic of Drawing History from Ancient Documents* (written in 1901, CSP CP 7.162–255). While his approach can be regarded as innovative at the time of its writing, it does not use the concept of abduction (at least not explicitly). It is therefore not relevant for my paper. Cf. Peirce (1966).

vehicles and what is signified" (Chandler 2007, 36). Peirce differentiates between the three modes: *symbol* (an arbitrary sign), *icon* (signs that resemble the signified) and *index* (a sign that is directly connected to the signified; Chandler 2007, 36).

As an example, "a weathercock is an index of the direction of the wind", as there is a "real connection" between the wind and the direction the weathercock is pointing to (CSP CP 2.286). Photographs or paintings, as another example, are icons, as the objects depicted in them—while not being the real thing—resemble the physical objects they represent (CSP CP 2.281). "The symbol is connected with its object by virtue of the idea of the symbol-using mind, without which no such connection would exist" (CSP CP 2.299).

14.5.3 Abduction, History and Geography

Maps, as Wood and Fels assess, are "complex super-signs" (Wood and Fels 1993, 132). Wirth analyses that a map hosts all three modes of a sign: Technical elements such as scale and projection are symbols, the relation between streets and cities is iconic and the naming of the streets and cities is indexical (Wirth 2007, 59). Such complex signs require training to decode, obviously. Cartography instils a grammar and syntax for maps on people, "a language of maps"; without such language, there would be a "cartographic illiteracy" (Schlögel 2009, 101). Signs are the result of an encoding process. The sign producer encodes and condenses (or: abstracts) a range of objects (the signified) into a sign; the reader of that sign than has to decode it to get the meaning of the sign, i.e. the signified (Wood and Fels 1993, 116). Most historians are not trained well to decode maps (or other geographical information), whereas geographers are usually not good at decoding narratives. Eminent German philosopher Immanuel Kant (1724–1804) put it this way: "History is a narrative, but geography is a description". For Kant, both history and geography relate to two phenomena in the world (space and time), and he regarded them as intertwined: Geography as the momentenous description of the world and history as a temporal one. For him, history is sequential geography: "History [. . .] differs from geography only in respect to space and time" (May 1970, 260).

I don't have a great affection for Kant's historiography, but his take on the relation of geography and history sheds some light on the significant difference of the two that has also some explanatory power for the dissatisfaction many historians have with Gistory: Historians and geographers don't think differently, they read differently.[22] Both groups, as I have pointed out above, have a different

[22]In the context of gistory, I think this issue—a lack of "Quellenkritik" (criticism of sources)—stems from the fact that gistorians often use only one source, or one type of source, such as census reports. It shouldn't be surprising that this defaulting to one kind of source doesn't hone one's skill in critically engaging with one's sources. On the other hand, it is painful to read how some historians try to validate all of their statements by a reference to a source. As Munz (1997, 868)

understanding of historic sources. Geographers tend to regard maps as *iconic*, sometimes *indexical*, signs representing an actual physical object (a landscape, etc.), whereas historians regard historic documents (including maps) as *symbolic signs* that represent the intentional statement its creator wanted to make or as *indexical signs* that are created by the historian interpreting the document.[23] This process by historians, discovering—postmodernists would say: imposing—indexical signs in history is referred to as "colligation" (Munslow 2000, 46). Historians colligate "hitherto disjointed and incongruous past events into meaningful historical sentences" (Munslow 2000, 47). This process—colligation—is an abductive one. *Evidence* (however defined) of an historic event is explored and hypotheses formulated to "explain an apparent anomaly" (Munslow 2000, 139). This is essentially the point Ginzburg made, as discussed above.

Historical research, by and large, is abductive in nature. Abduction, however, as Peirce pointed out is just one stage in an inferential process. Ginzburg and others have therefore regarded the clue principle in history as one stage in historical research. Ginzburg, above all others, has perceived the clue paradigm as a bridge between adverse historical schools such as postmodernism and positivism. While postmodernist philosophers of history such as Alan Munslow use the abductive nature of history to maintain that "history's conclusions must be provisional" regardless of any "evidence" (Munslow 2000, 140), tracking, in Sybille Krämer's words, reconciles these camps by providing evidence for a historical construction (Krämer 2007a, 173). I will come to this issue in historiography and how GIS fits into it further below. Before that, however, it would be a good idea to point out the difference between abduction and tracking. As I will outline later, while a number of tools and techniques are quite helpful for abductive reasoning, I regard GIS as especially qualified for tracking.

14.5.4 Tracking and Abduction

Tracks, as Stegmaier rightfully observes, almost always refer to past events (Stegmaier 2007, 82). Or as Sybille Krämer remarks: "A clue becomes a track when at the time of reading it the clue is showing something that has irreversibly passed", but only when the clues are left behind unintentionally. What Peirce pointed out was that the clues being found are transformed/abducted (by experience, i.e. prior knowledge) into unambiguous conclusions (the ones that are the most likely explanations). For example, we can infer the presence of a mouse in our house from "squeaking noises" and "vanishing cheese" (and other clues such as mouse

rightfully points out, this enterprise becomes circular; at one point, the historian has to leave the "proofs" behind and start making abductive inferences (i.e. start "guessing", see above).

[23]Cf. Wirth (2007, 69). I am well aware that not all geographers see maps this way, but from my experience, the majority of geographers involved in Historical GIS/Gistory do.

droppings; Richmond 1999, 86). While there might be other explanations for these clues, like them being a prank, once we gathered enough information that coincides with our past experience (having heard mouse squeaks before, having read that mice love cheese, etc.), it is the most plausible one.

Tracks are therefore not indexical signs per se, but have a temporal edge. A good example that Krämer uses is—again—that of a hunter: If a hunter encounters a track, time is of the essence. He or she may correctly identify the track as the game being sought, but if the track is days old (or even more), that inference is pointless. Only the temporal dimension makes the inference useful, it is only the temporal dimension that turns a track into an indexical sign. That is, if the spoor is recognised as sufficiently fresh, it is transformed into an indexical sign by treating it as if the game was temporally and geographically close enough to the hunter to be usable (Krämer 2007a, 165).

Abduction is thus not equivalent to tracking, but tracking is transformed into a sequence of abductive inferences. So when abduction can be likened to "guessing", tracking can be regarded as "the art of intelligent guessing" (Krämer 2007b, 21). It adds an element of speculation and creativity to the process of abductive reasoning. Peirce himself admitted that abduction would describe the action of "guessing" (CSP CP 5.173). He explained this phenomenon as being based on intuition and instinct, i.e. experience and common sense. I have mentioned the necessity of past experience to decode the signs found on a track. The right balance of speculation and drawing on experience is essential. Meaning: While a certain level of prefiguration is required to make sense of signs, too much of it turns into preoccupation and hinders one to infer the right conclusions from the clues found.[24] As Kaplan argues, discoveries are guided by prior experience and one's (figurative) position in the field (i.e. the subject); we look for things because "we remember having it seen there" and we also dismiss looking in other places because "we happen to be in the place already [or] others are looking elsewhere, and so on" (Kaplan 1973, 18). This makes sense in both our hunting metaphor as well as in scientific practice. If I see an old spoor on a hunting trip, I will most likely not pursue it on that day, for it is not leading me to game currently in the area, but it informs me of game having traversed the area, which means it may do so frequently, and thus it would be a good spot to explore in the future. Likewise, if other hunters occupy the other hunting grounds in the area already, there's not much point joining them. In research, the situation is similar. I may make an observation that does not yield much at the time, but could prove significant later. In the same vein, if other researchers work on a certain topic, it is probably wiser to focus on other ones.

---

[24]Krämer (2007b, 21). What Peirce meant, though, was probably "intuition" rather than "guessing"; as Kaplan (1973, 14) explains the difference: "There is surely a basic difference between intuition and guesswork—between the intuition of the great creative genius or even of the ordinary experienced scientific worker, and the complete novice's blind, blundering guesswork or mechanical trial and error".

Tracking is—to pick up the point about exploration from above—what Stebbins called "concatenated exploration": finding clues consecutively and linking them together in a chain, eventually leading to a cumulatively generated theory (Stebbins 2001, 12). The main difference, though, between abduction and tracking is the former's occupation with intuition to arrive at inferences, whereas the latter also introduces an element of speculation into the mix. As Peter Carruthers writes, other than in "ideal conditions [. . .] a mere capacity to recognize and follow an animal's spoor will be by no means sufficient to find it" (Carruthers 2002, 80). Rather: "hunters will often need to develop speculative hypotheses concerning the likely causes of the few signs available to them" (Carruthers 2002, 81). Other than abduction, which usually operates on a fixed set of observations, tracking may very well encounter highly dynamic clues, i.e. clues that change along the track. A spoor may change, may have been diminished (e.g. the game running over different type of soils that allow less clues) or may expose more clues than where the spoor started. To deal with this dynamism, speculation helps to fill the gaps in the clues at hand. While speculation is part of every historian's work to some extent, until recently there was little attention given to a more systematic approach of speculative history, often subsumed as "virtual history".

14.5.5 Virtual History

The rather new field of "virtual history" can be regarded as some kind of descendant of the postmodern branch of history—although some of its most prominent proponents like Niall Ferguson are far from subscribing to the underlying political and philosophical implications of postmodernism, such as relativism. Postmodernism, anyway, "with its blurring of the boundaries between fact and fiction, its privileging of 'other' or alternate voices" (Rosenfeld 2002, 92) has paved the way for "virtual", "counterfactual" or "alternate" history.

Virtual history is therefore not about Ranke's "how it actually was" paradigm, but rather about what German historian Helmut Fleischer put as "asking on how and what it might have been" (Fleischer 1998, 417) or "history as it might have, must have, should have been" (Landes 1994, 637). Virtual history is probably best known in its vulgar variant—that is, alternate history stories as seen in film and novels that depict an alternate course of history and what the present would look like had, for example, Nazi Germany won the Second World War (this seems to the most popular topic in these alternate history stories). This form of virtual history first and foremost serves the purpose to reflect on present political debates, rather than exploring the past (Rosenfeld 2002, 103). While this angle of virtual history surely has its own allure, the point I am getting at is the use of it to speculate about and the comparison of historic narratives.

Niall Ferguson rightly points out that it is not the imaginative or fictional branch of virtual history that should interest us but rather the one that is "designed to test hypotheses by (supposedly) empirical means, which eschew imagination in favour of computation" (Ferguson 1997, 18). It is therefore a method to test and compare different historic narratives and hypotheses. Ferguson argues that "counterfactuals" are an essential learning technique for the human mind (Ferguson 1997, 2). The human brain, if you follow this theory, learns chiefly by trial and error and by inferences made by that. Counterfactuals are part of this approach, allowing one to speculate about possible outcomes of one's actions (and then picking the most rewarding one) or—in our case of historical studies—to speculate about possible hypotheses and their plausibility. This speculation is primarily designed to explore the historic narratives and not to find the "true" one out of many rivalling ones.[25]

Staley has already hinted at the potential of GIS to function as a "geographically referenced counterfactual" for history (Staley 2003, 125). I would like to take this point further by arguing that GIS is a tool to employ Ginzburg's clue paradigm (as I outlined above) and that counterfactuals fall into the same category as this paradigm. Bonnett has discussed virtual history in the context of "thought experiments" to test "the plausibility of the candidate hypothesis" by checking whether the outcome of the experiment "maps, or nearly maps, with the historical record" (Bonnett 2007, 3). Given that we frequently encounter concurrent—often contradicting—and anachronistic data in historical records, a virtual history approach to GIS in historical research would give us a tool not only to test hypotheses but also to create new hypotheses by exploring and speculating about the data stored in the GIS. Bonnett regards "thought experiments" as synonymous to abductive reasoning, a principle that Ginzburg's clue paradigm is also founded on, as I pointed out above.

14.5.6 Tracking and Counterfactuals

This is to say that virtual history is not entirely equivalent to tracking but follows the same basic principle. Just like a hunter infers a string of properties of the game from a given track through its characteristics (i.e. inferring the signified from the signs), so does the historian extract characteristics from the sources. To infer from an animal's track its physical properties works by comparison and contrasting, i.e. the track being read is compared to the ones previously encountered, and then the difference between the two is used to make inferences about the animal in question. Many of the steps involved in this process, however, depend on facts not so readily available, and hunters have to draw on both intuition and experience to make sense of the track and develop a strategy to exploit the inferences (i.e. to hunt the game down). In other words, hunters employ what Mey and Weber have called "contrastive reasoning" (Mey and Weber 2003, 35). Thought experiments and counterfactuals are used to arrive at "weighted explanations of contrasts" in this respect (Mey and Weber 2003, 38).

[25] Ferguson regards counterfactuals as a way to defy determinism rather than to elevate it (Ferguson 1997, 55ff).

These "weighted explanations" are perfectly in line with Peirce's notion that "Abduction merely suggests that something *may be*" (CSP CP 5.171). The main difference being—and this forms the link between counterfactuals and tracking—is the circumstance that abduction tries to arrive at *one* explanation, whereas tracking may have many of them, as mentioned above. The process of tracking—in comparison to abduction—is thus to continually come up with "weighted explanations" of the "contrasts" (i.e. the difference in tracks/signs/clues encountered and those memorised) until a satisfying state is reached: the game being hunted down, in this example. Within this process, possible explanations will be eliminated, and sometimes the hunter will make a counterfactual assumption to imagine alternative patterns of behaviour or simply because the assumption will eliminate more potential explanations than other hypotheses would, thereby reducing complexity and speeding up the decision making (Kaplan 1973, 17).

Outside the domain of hunting, things might be a bit more complicated. In detective work, many more variables are likely to be present than just clues. In crime investigation, other circumstances such as possible beneficiaries (i.e. who would benefit from the crime to limit the number of suspects, the "cui bono" principle) play an important role. Forensics in crime investigation makes heavy use of model making, both physical models and computer models. One example is to use "toy cars to reconstruct an auto accident" (Shelley 2003, 640). Computers have been used, among other things, to calculate and reconstruct the shooting angle from projectiles found in walls and such at crime scenes, for example. While the reconstruction obviously doesn't solve the crime instantly, it profoundly helps detectives to do so by filling gaps in the evidence.

In historical scholarship, this situation is not unknown. Historical sources are inherently full of gaps, and the historian's argument is based on how he or she is able to fill them. In literature theory, this involves the process which is called "overreading": "we find in narratives qualities, motives, moods, ideas, judgments, even events for which there is no direct evidence in the discourse" (Abbot 2002, 82). While leaving the gaps intact is an attractive feature of literature to leave something to the imagination of the reader, in history, this is usually unsatisfying. The interpretation of an historic narrative is therefore based on some evidence and some overreading, i.e. interpretation by the historian.[26] Overreading is inevitable in historical scholarship no matter what the individual historian might think about the "evidence" at his or her fingertips. The "line between inference and imagination is [...] regularly crossed by the historian" (Munslow 2000, 124)—regardless whether historians admit it or not.

One approach to tackle gaps in narratives is *intentional* overreading (Abbot 2002, 93). Virtual history is such intentional overreading, while those historians not subscribing to virtual history overread too, yet just within a different scope. Taking these strands together—tracking, clues, abduction, overreading, counterfactuals—

[26]"In this regard, narrative interpretation is no different from analysis in most other fields in which meaning must be culled from data" (Abbot 2002, 90).

we should be able to discern a pattern here. Given that all historical scholarship involves abductive reasoning and overreading, what would constitute a most useful tool for intentional overreading in history? Given that tracks involve geography as well as time, as pointed out above, GIS seems to be made for such an enterprise. Tracks have a place and reading a track means "to reconstruct the action that led to the track as a narration" (Krämer 2007b, 17). GIS not only helps to achieve such reconstruction by filling the gaps in the sources but also by allowing the historian to create counterfactuals and use them for contrastive reasoning.

14.5.7 Tracking and GIS

In the context of Gistory, the principle of tracking seems therefore almost self-evident. Tracking is finding clues over space and time, and so is Gistory. Just as an animal leaves clues of itself in the environment, which the hunter tracks, so do humans leave clues about their activities in records, documents and artefacts. What would be more obvious than using GIS to track those human activities over space and time?

Expanding on the distinction I made earlier between historical GIS and gistory, it may appear that there is no difference between the two in regard to tracking. HGIS simply store all available clues about a (usually) very broad range of topics, while gistory approaches also have to store a good deal of data, albeit a bit more specific to the historic question being explored. It has been argued that historical GIS are simply about storing the "raw" data and that the historian can then work with it, as with any other archive. I strongly object to this notion of neutrality of data. Storing data in a database—however sophisticated—includes the stage of data modelling. Data modelling involves interpreting the data, i.e. there is no such thing as "raw" data in a database. Or put differently: The act of modelling data embodies the early stages of the tracking—the interpretation of the clues—and there is no way of separating the two.

It is obvious that the data model for the same data will differ with every historian—or even for the same historian at different times, as more expertise and different research questions will alter the model significantly. This, however, is in no way different from what historians do anyway. The same source will hardly be read and interpreted the same way by different historians, and a review of the same sources by the same historian some years later will most likely not yield the same interpretation.

Consequently, tracking starts from early on in the research project, and the data model will basically depend on how the sources "speak" to the historian. Ginzburg's clue paradigm postulates that historic narratives reconstruct clues about the events reported in them. While the narrative as such might be highly impartial, they still offer some insight into the historic context, though. In another landmark publication of Ginzburg, he uses a "vision" metaphor to explain the differences in historical methods. While positivists take sources as "open windows" into the past, postmodernists see them as "fences obstructing vision". Rather, Ginzburg argues, sources are like "distorting mirrors" from which the historian both constructs and reconstructs the historic context of past events (Ginzburg 1999, 25).

As I have argued above, on top of the narrative that is the historic source which speaks to the historian, a GIS offers yet another narrative. By overlaying different maps, for example, historians can try to analyse this distortion Ginzburg spoke of and thereby reconstruct the historic event, and synthesise (i.e. construct) the historic context. Furthermore, the historian does not remain a consumer/reader of maps but also becomes a map producer. The use of GIS more than often does not restrict itself to displaying scanned historic maps, but it generates new maps out of the combination of various data layers. A scanned historic map can be combined with demographic data or topological features to check different sources in context. The maps so produced do not have to meet the highest cartographic standards but can rather function as "scratchpad" for the historian to explore the data and to create and test hypotheses. Those maps are to be used in a contrastive reasoning by overlaying data sets and scanned maps and visualisations. These layers do not have to refer to actual data sets, such as tables from historic census reports, but can be entirely made up—under the constraints of what passes as sensible in the research agenda, of course. That is to say that the GIS layers and the (intermediate) maps produced from them can be created (at least in part) by counterfactual data, data that the historian has either constructed from thought experiments, and may be simply modifications of existing layers. Or it could be material that has a different historic context, for example from different historic periods. The significant point is that the historian not only decodes signs, as he or she would when reading/interpreting historic maps, but the historian would also be empowered to encode signs. He or she could take what was formerly the *signified* (i.e. the data or the map that symbolised a person, a landscape or a social process to name but a few possibilities) and turn it into a *signifier* (i.e. rearrange or reinterpret symbols that others have created to thus create your own symbol).

This is not about asking for arbitrary map making, like computer game designers are creating maps for virtual worlds, but to reconstruct tracks from both sources and the inferences and speculations induced by them; or rather: abduced through them. Just as I have outlined above by discussing the abductive nature of historical research and Ginzburg's "clue paradigm" and history as tracking, these maps produced via GIS reconstruct the actions that led to the clues discovered by the historian as a narrative process.

14.5.8 The Semiotics of Space of Ancient Rome

Perhaps a short example would shed some light on the idea behind this philosophical venture. Between 2005 and 2007, I was involved in a research and teaching project on the border installations the Ancient Romans erected all around their empire. These installations (ramparts, ditches, walls) were dubbed "Limes" by nineteenth century German historians, the Latin word for "limit" or "border". The mere fact that the Romans never called these borders "Limes" themselves was reason enough to doubt many established theses about its function already. As it turns out, the introduction of the term reflects German notions of nationality and nationhood in the nineteenth century rather than contemporary Roman spatial models.

After struggling to make sense of the geographical setup (i.e. the infrastructure) of the Limes, where geo-statistical analyses of buildings and archaeological findings would not sufficiently clear up apparent contradictions (i.e. that the Romans went to great length to seal off their territory, when little to no opposition could be found on the other side of it), we turned the GIS-driven analysis on its head. Rather than constructing an explanation from available data and relying on established quantitative methods to do so, we started from historical spatial thinking of the Romans to interrogate the data; i.e. we started from the premise that a historic spatial infrastructure such as the Limes and related buildings are *signs* of Roman mentality. We then used geometric and statistical analyses to pursue our methodology. The quantitative approach illustrated that the infrastructure made sense locally but overcompensated within a larger geographical context. We found that there were in fact local reasons to protect such an area (e.g. an area with many granaries in it). Yet this alone could not explain the rationale and effort to set up walls and turrets in areas where economic interests did not exist. It did not explain why Limes in one part in Germany were built in an almost perfect straight line for c. 80 km and ignore landscape features that could be militarily exploited in its course.[27] By analysing the data in GIS, or simply by visualising the data with unconventional methods, we step by step discovered that the main drive for the Limes was not military or economic in nature but rather cultural: It delineated a "civilised" Roman society from the "barbarians" world outside the Limes. It introduced for the first time in human history an artificial, arbitrary border where hitherto only natural borders were used to delimit one's territory. We referred to this approach as a "semiotics of space", as we were more concerned with the *signs* that the Romans built in the landscape and not with how many Limes they built (as the usual statistical approach would have been employed in so many Gistoric projects).[28]

All of the methods we employed in this research were already available in the GIS software; it was just that we used them in a manner one would not necessarily find in the manuals. The maps thus produced would make most cartographers and geographers tear their hair out, but they were not intended to visualise our findings, but rather functioned as a kind of sketchpad for us. The maps informed us about alternative scenarios, helped us memorise and discuss them and created the conditions for the "contrastive reasoning" I discussed above.

---

[27] Geodetic surveys have found that the Limes strayed only maximally 1.5 m from an ideal straight line.

[28] See Lünen and Moschek (2011) for details and also the contribution of Mares and Moschek in this book. That the Limes is a signpost of Roman mentality was Moschek's main point in his doctoral dissertation, and from that point of departure, I blended this with a GIS approach.

What was accomplished in that paper was to a certain extent relevant to the terms "public symbols" and "fields of care" from Tuan (1979) to which I referred to earlier. We regarded the Roman-built Limes as both "public symbol" (meaning that it is immediately visible, i.e. with military and economic utility) and as "field of care" (having concealed meaning that is only discovered by experience, i.e. knowledge of and abductive reasoning about Roman culture; cf. Tuan 1979, 412). Swedish philosopher of geography Gunnar Olsson is also concerned with this "mapping of the invisible". He asserts that "without imagination there would never be any maps" (Olsson 2007, 120). He poignantly assesses that "cartographical reasoning" is a "methodological two-step: *First* the investigated phenomenon is translated into the graven image of a map, *then* that idol is blessed with a set of seemingly meaningful interpretations" (Olsson 2007, 138). This pattern is all too prevalent in both HGIS and gistory: Data is collected and collated, processed, stored and eventually mapped. Then, in a second step, the maps are interpreted, taking the underlying data and the map it produces for granted and thus disconnected from the mapping process.[29] However, our tools determine our work, and the mapping process can therefore not be regarded as independent from the interpretation of the map.[30] GIS, however, empowers its user to reason and interpret the underlying data and the map that is constructed from it while both are still being worked on. GIS work is a permanent interpretative process, and its user will engage with his or her data at every corner.

I think it is this property of GIS that makes it an invaluable tool for *tracking* (as in Ginzburg's "clue paradigm"): Historians can utilise GIS to *arrange* and *elicit* rather than *analyse* historical narratives to make them perceive things in a different light. GIS should thus not—or at least not only—be a tool to visualise outcomes to an audience, but rather for constructing new narratives in order for historians to constantly reassess their own research agenda. In this light, historians don't have to become GIS experts or programmers, but rather work like (to borrow a term from anthropology) *bricoleurs*, who work "with signs, constructing new arrangements by adopting existing signifieds as signifiers [...]" (Lévi-Strauss 1968 [1962], 62).

14.6 Conclusion: The Historian as Bricoleur

French anthropologist Claude Lévi-Strauss (1908–2009) introduced the term "bricoleur" into science studies. The term as such is somewhat difficult to render into English but would probably best translate as "jack-of-all-trades" or "handyman".[31] The traditional meaning of the word referred to deflections in ball games or game movements in hunting (i.e. the term refers to a kind of movement that is not

---

[29] See the interview with Emmauel Le Roy Ladurie in this book and his cooperation with Jacques Bertin.

[30] See the interview with Gunnar Olsson in this book.

[31] Brick (2002, 100) gives "handyman" or "DIY man" as translation for bricoleur and "tinkering about" and "makeshift repair" for "bricolage".

straight to goal but with a detour, such as "to avoid an obstacle"; Lévi-Strauss (1968 [1962], 16)). "Bricoleur" is someone who "uses devious means compared to those of a craftsman" (Lévi-Strauss 1968 [1962], 17). Lévi-Strauss sets the bricoleur in juxtaposition to the engineer. The latter approaches a technical problem in a very systematic manner, creating the tools he or she needs to accomplish a task. The bricoleur, on the other hand, takes the tools and materials readily available and makes the best use of them (Lévi-Strauss 1968 [1962], 18). He (or she)—the bricoleur—is therefore very creative in the use of existing tools, applying them to tasks for which they were not originally designed for,[32] as his or her tools do not "have only one definite and determinate use" (Lévi-Strauss 1968 [1962], 18).

The engineer and the bricoleur differ in another significant respect as well, according to Levi-Strauss: "the engineer works by means of concepts and the bricoleur by means of signs" (Lévi-Strauss 1968 [1962], 20). The engineer assesses the problem, analyses the difficulties and considers the tools and materials required, i.e. prior to any actual work, the engineer has made a plan in all details on how to tackle the problem. The bricoleur picks the tools as the problems come. He or she gets into a "dialogue" with his or her "set of tools" and "before choosing between them [has] to index the possible answers which the whole set can offer to this problem" (Lévi-Strauss 1968 [1962], 18).

Historian of technology David Edgerton makes a similar point when he speaks of the "creolisation of technology" that car repairers in African countries have developed. Being cut off from the Western technical maintenance networks these "tinkerers" had to develop unorthodox methods to repair cars "using local materials". By doing this, they not simply managed to keep the cars going, but "they transformed cars" (Edgerton 2008, 84–85). The methods, tools and materials would make the European engineers who designed the cars cringe. Yet, in the local context, they make perfect sense. It is the same with Gistory. Cartographers, geographers and GI scientists may snub the way historians employ maps and GIS (or spatial concepts in general), but that's what historians feel works best for them.

Sherry Turkle has adopted the principle of the bricoleur to computers: "Computers [...] make possible a theoretical tinkering similar to 'bricolage'" (Turkle 1999, 543). For her, "bricolage" embodies "the process by which individuals and cultures use the objects around them to reconfigure the boundaries of their cognitive categories" (Turkle 1999, 543). It is this attitude I see as necessary for historians to appropriate in order to make any headway with GIS—or computers in general—in historical scholarship.

What this encompasses should be clear: Historians have to leave the well-trodden path set out by GIScience and chart their own territory. This does not to be done from scratch. Geography and GIScience are a great inspiration for

[32]This is by and large the definition of a "hacker": the creative use of technology. Hacking, however, also implies a joyous and playful use of technology, something that would not necessarily apply to the bricoleur. I have used the concept of hackers in my doctoral dissertation to explain why some engineers were more successful at developing pressure suits in aviation in the 1930s than others. Cf. Lünen (2010b, 251ff).

the gistorian—but nothing more. According to Levi-Strauss, the bricoleur "creates improvised structures by appropriating pre-existing materials" (Chandler 2007, 205). The mere fact that a gistorian's work with GIS would qualify as provisional and not up to cartographic or geographic standards should not deter, but encourage him or her.

All too often, scholars in Historical GIS impersonate the engineer, deliberating in pedantic detail the construction plan of their project, creating new tools to fit the purpose (i.e. develop new HGIS) and remaining within the confines of the methodological framework their respective subject sets out. Someone likened the bricoleur to the fictional character of MacGyver, the daredevil detective from the TV series in which the main character is best known for his bewildering construction of tools out of household goods and other common components. In this vein, I ask for GIS-aficionados to become the MacGyvers of history, i.e. I ask for a sort of eclecticism that uses the tools as it finds them and combines these tools in an unconventional manner. It should not matter which school you subscribe to—postmodernism or positivism or something else—or whether you deem the current GIS software packages up to the task, for computers are just tools and bear no originality: "Thus according to Peirce's normative definition of inference a computing machine fails to infer in two respects, conscious approval and originality, both of which are essential to abductive reasoning" (Burks 1946, 305). We therefore don't need a new GIS technology, or at least not per se, but rather to make better use of the ones we already have.[33]

Jerome de Groot has made this point in his *Hacking history: Google Earth* chapter (de Groot 2009, 98ff). He regards open source mapping as tools that can help to "escape the gatekeepers of cartographic knowledge" and give "users free rein to create and control their own sense of space".[34] De Groot argues that this "potential for disruption inscribed within [GoogleEarth] allows historical 'fact' to become another set of data to be manipulated in whichever way the end-user wishes" (de Groot 2009, 99). I do not share his affection for GoogleEarth, yet I endorse his view that historians should neither accept one certain product as *the* GIS nor that one certain discipline should be adhered to for how to make use of GIS-ish technology.

It could be argued that I ask for an *appropriation* of GIS technology by historians, i.e. historians learning the tools and trade of GISystems and science to adapt it to their subjects. I go but further and ask for an *expropriation* of GIS technology, to drop "historical GIS" in favour of "gistory". GIScience simply cannot answer, let alone satisfy, the various needs and perspectives of historians. It is thus about time historians usurp normative power and apply GIS to historical research at their own terms, rather than what GIScience and geography proscribe.

---

[33] I am aware that this sentiment puts me in opposition to other gistorians, such as Wachowicz and Owens in this volume; see also (Owens 2007). But as a free market advocate, I leave it to the reader to decide which argument is more appealing.

[34] de Groot (2009, 98). For the concept of "gatekeepers" in the use of GIS, etc, see Mark Palmer's contribution in this book.

Rather than a visualisation tool, GIS should be used as a painting tool: a tool to creatively engage with one's sources. As Gunnar Olsson argues: "GIS stands naked before us, shamefully parading as a game of ontological transformations in which theory-laden observations are translated first into patches of color, then into strings of words, finally into purposeful action. Picture becomes story, 'is' turns to 'ought'" (Olsson 1999, 141). This is the decision to be made by historians: Will they remain in the "painting by numbers" domain, or will they be able to develop into true GIS artists, abandoning the imitation game and transform GIS into a genuine vehicle for historic inquiry?

References

Abbot HP (2002) Cambridge introduction to narrative. Cambridge University Press, Cambridge

Andrienko N, Andrienko G (2006) Exploratory analysis of spatial and temporal data. A systematic approach. Springer, Berlin

Aucott P, Lünen Av, Southall H (2009) Exposing the history of Europe: the creation of a structure to enable time-spatial searching of historical resources within a European framework. OCLC Syst Serv 25(4):270–286

Bentley M (1999) Modern historiography: an introduction. Routledge, New York/London

Blunt A (2003) Geography and the humanities tradition. In: Holloway SL, Rice SP, Valentine G (eds) Key concepts in geography. Sage, London, pp 73–91

Bonnett J (2007) Abductive reasoning, A-Life, and the historian's craft: one scenario for the future of history and computing. http://www.arts-humanities.net/system/files/reasoning.pdf. Accessed 04 Aug 2011

Brick M (ed) (2002) Collins/Robert French-English English-French dictionary. Unabridged. Harper Collins/Dictionaires Le Robert, Glasgow/Paris

Bullough VL (1966/1967) The computer and the historian – Some tentative beginnings. Comput Humanit 1:61–64

Burke P (2009 [2005]) History and social theory, reprint edn. Polity, Cambridge

Burks AW (1946) Peirce's theory of abduction. Philos Sci 13(4):301–306

Carr E (1990) What is history? Reprint of the 2nd edn. Penguin, London. First published in 1961, 2nd edn from 1987

Carruthers P (2002) The roots of scientific reasoning: infancy, modularity and the art of tracking. In: Stich SP, Siegal M, Carruthers P (eds) The cognitive basis of science. Cambridge University Press, New York, pp 73–95

Castree N (2003) Place: connections and boundaries in an interdependent world. In: Holloway SL, Rice SP, Valentine G (eds) Key concepts in geography. Sage, London, pp 165–185

Chandler D (2007) Semiotics: the basics, 2nd edn. Routledge, London/New York

Daintith J, Wright E (2008) A dictionary of computing, 6th edn. Oxford University Press, Oxford

Davies S (2003) Empiricism and history. Palgrave, Basingstoke

de Groot J (2009) Consuming history: historians and heritage in contemporary popular culture. Routledge, London

Döring J, Thielmann T (2008a) Einleitung: was lesen wir im Raume? Der Spatial Turn und das geheime Wissen der Geographen. In: Döring J, Thielmann T (eds) Spatial turn: Das Raumparadigma in den Kultur- und Sozialwissenschaften. Transcript, Bielefeld, pp 7–45

Döring J, Thielmann T (eds) (2008b) Spatial turn: Das Raumparadigma in den Kultur- und Sozialwissenschaften. Transcript, Bielefeld

Edgerton D (2008) The shock of the old: technology and global history since 1900. Profile, London

Ferguson N (1997) Introduction. Virtual history: towards a 'chaotic' theory of the past. In: Ferguson N (ed) Virtual history: alternatives and counterfactuals. Picador, London, pp 1–90
Fleischer H (1998) Mit der Vergangenheit umgehen. Prolegomena zu einer Analytik des Geschichtsbewußtseins. Leviathan 18(Sonderheft):409–432
Ginzburg C (1976) The cheese and the worms: the cosmos of a sixteenth-century miller. Routledge, London
Ginzburg C (1989) Clues, myths, and the historical method. John Hopkins University Press, Baltimore
Ginzburg C (1999) History, rhetoric, and proof. University Press of New England, Hanover
Gove PB (1976) Webster's third new international dictionary of the english language unabridged with seven language dictionary. G. & C. Merriam, London
Harman GH (1965) The inference to the best explanation. Philos Rev 74(1):88–95
Heffernan M (2003) Histories of geography. In: Holloway SL, Rice SP, Valentine G (eds) Key concepts in geography. Sage, London, pp 3–22
Hill L (2006) Georeferencing: the geographic associations of information. MIT, Cambridge
Holloway SL, Rice SP, Valentine G (eds) (2003) Key concepts in geography. Sage, London
Iggers GG (1997) Historiography in the twentieth century: from scientific objectivity to the postmodern challenge. Wesleyan University Press, Middletown
Kaplan ADH (1973) The conduct of inquiry: methodology for behavioral science. Intertext, Aylesbury
Krämer S (2007a) Immanenz und Tranzendenz der Spur: Über das epistomologische Doppelleben der Spur. In: Krämer S, Kogge W, Grube G (eds) Spur. Spurenlesen als Orientierungstechnik und Wissenskunst. Suhrkamp, Frankfurt/M, pp 155–181
Krämer S (2007b) Was also ist eine Spur? Und worin besteht ihre epistomologische Rolle? Eine Bestandsaufnahme. In: Krämer S, Kogge W, Grube G (eds) Spur. Spurenlesen als Orientierungstechnik und Wissenskunst. Suhrkamp, Frankfurt/M, pp 11–33
Krämer S, Kogge W, Grube G (eds) (2007) Spur. Spurenlesen als Orientierungstechnik und Wissenskunst. Suhrkamp, Frankfurt/M
Landes DS (1994) What room for accident in history? Explaining big changes by small events. Econ Hist Rev New Ser 47(4):637–656
Levi G (1991) On microhistory. In: Burke P (ed) New perspectives on historical writing. Polity, Cambridge, pp 93–113
Lévi-Strauss C (1968 [1962]) The savage mind (La Pensée Sauvage), 2nd edn. Weidenfeld and Nicolson, London
Lünen Av (2010a) Historische GIS als Narrative? In: Mitterbauer H, Scherke K, Müller S (eds) Kulturwissenschaftliches Jahrbuch Moderne, vol 5. StudienVerlag, Vienna, pp 221–225
Lünen Av (2010b) Under the waves, above the clouds: a history of the pressure suit. Ph.D. thesis, Technical University Darmstadt, Darmstadt, Germany. http://tuprints.ulb.tu-darmstadt.de/2103/. Abridged version of the submitted version from 2007. Accessed 10 May 2012
Lünen Av, Moschek W (2011) Without limits: ancient history and GIS. In: Dear M, Ketchum J, Luria S, Richardson D (eds) Geohumanities: art, history, text at the edge of place. Routledge, London
Marble DF (1990) The potential methodological impact of geographic information systems on the social sciences. In: Allen KMS, Green SW, Zubrow EBW (eds) Interpreting space: GIS and archaeology. Taylor & Francis, pp 9–21
May J (1970) Kant's concept of geography: its relation to recent geographical thought. University of Toronto Press, Toronto
Mey TD, Weber E (2003) Explanation and thought experiments in history. Hist Theory 42(1):28–38
Middell M (2008) Der Spatial Turn und das Interesse an der Globalisierung in der Geschichtswissenschaft. In: Döring J, Thielmann T (eds) Spatial turn: Das Raumparadigma in den Kultur- und Sozialwissenschaften. Transcript, Bielefeld, pp 103–124
Munslow A (1997) Deconstructing history. Routledge, London
Munslow A (2000) The Routledge companion to historical studies. Routledge, London

Munz P (1997) The historical narrative. In: Bentley M (ed) Companion to historiography. Routledge, London/NewYork, pp 851–872
Olsson G (1999) Constellations. Sist Terra 8(1–3):140–144
Olsson G (2007) Abysmal: a critique of cartographic reason. University of Chicago Press, Chicago
Owens J (2007) What historians want from GIS. ArcNews 29:4–6
Peirce CS (1960a) Collected papers of Charles Sanders Peirce. Principles of philosophy, vol 1; Elements of logic, vol 2, new edn. Belknap Press of Harvard University Press, Cambridge, MA
Peirce CS (1960b) Collected papers of Charles Sanders Peirce. Pragmatism and pragmaticism, vol 5; Scientific metaphysics, vol 6, new edn. Belknap Press of Harvard University Press, Cambridge, MA
Peirce CS (1966) Collected papers of Charles Sanders Peirce. Science and philosophy, vols 7/8, 2nd edn. Havard University Press, Cambridge, MA
Piltz E (2008) 'Trägheit des Raums'. Ferdnand Braudel und die Spatial Stories der Geschichtswissenschaft. In: Döring J, Thielmann T (eds) Spatial turn: Das Raumparadigma in den Kultur- und Sozialwissenschaften. Transcript, Bielefeld, pp 75–102
Richmond A (1999) Between abduction and the deep blue sea. Philos Quart 49(194):86–91
Rosenfeld G (2002) Why do we ask 'what if?' Reflections on the function of alternate history. Hist Theory 41(4):90–103
Schlögel K (2009) Im Raume lesen wir die Zeit. Über Zivilisationsgeschichte und Geopolitik, 3rd edn. Fischer, Frankfurt/M
Shelley C (2003) Lorenzo Magnani: abduction, reason, and science: processes of discovery and explanation. Philos Sci 70(3):639–643
Sombart N (1992) Nachrichten aus Ascona. Auf dem Wege zu einer kulturwissenschaftlichen Hermeneutik. In: Prigge W (ed) Städtische Intellektuelle. Urbane Millieus im 20. Jahrhundert, Fischer, Frankfurt/M, pp 107–117
Southall H, von Lunen A, Aucott P (2009) On the organisation of geographical knowledge: data models for gazetteers and historical GIS. In: 5th IEEE international conference on e-science workshops, pp 162–166. DOI 10.1109/ESCIW.2009.5407970
Staley DJ (2003) Computers, visualization, and history. How new technology will transform our understanding of the past. Sharpe, London
Stebbins RA (2001) Exploratory research in the social sciences. Sage, London
Stegmaier W (2007) Anhaltspunkte. Spuren zur Orientierung. In: Krämer S, Kogge W, Grube G (eds) Spur. Spurenlesen als Orientierungstechnik und Wissenskunst. Suhrkamp, Frankfurt/M, pp 82–94
Tuan YF (1979) Space and place: humanistic perspective. In: Gale S, Olsson G (eds) Philosophy in geography. Reidel, Dordrecht/Boston/London, pp 387–427
Tukey JW (1977) Exploratory data analysis. Wesley, Reading, MA
Turkle S (1999) What are we thinking about when we are thinking about computers? (1995, abridged 1998). In: Biagioli M (ed) The science studies reader. Routledge, London, pp 543–552
Windschuttle K (1997) The killing of history: how literary critics and social theorists are murdering our past. The Free Press, New York
Wirth U (1995) Abduktion und ihre Anwendung. Zeitschrift für Semiotik 17(3/4):405–424
Wirth U (2007) Zwischen genuiner und degenerierter Indexikalität: Eine Peircesche Perspektive auf Derridas und Freuds Spurbegriff. In: Krämer S, Kogge W, Grube G (eds) Spur. Spurenlesen als Orientierungstechnik und Wissenskunst. Suhrkamp, Frankfurt/M, pp 55–81
Wittgenstein J (1978) Remarks on the foundation of mathematics, 3rd edn. Blackwell, Oxford
Wood D, Fels J (1993) The power of maps. Routledge, London
Yu CH (2011) Abduction? Deduction? Induction? Is there a logic of exploratory data analysis? http://www.creative-wisdom.com/pub/Peirce/Logic_of_EDA.html. Paper presented at the annual meeting of American educational research association, New Orleans, Louisiana, April, 1994. Last accessed 10 Aug 2011

Index

A. von Lünen and C. Travis (eds.), *History and GIS: Epistemologies, Considerations and Reflections*, DOI 10.1007/978-94-007-5009-8,

FSC
www.fsc.org
MIX
Papier aus verantwortungsvollen Quellen
Paper from responsible sources
FSC® C105338

If you have any concerns about our products,
you can contact us on
ProductSafety@springernature.com

In case Publisher is established outside the EU,
the EU authorized representative is:
Springer Nature Customer Service Center GmbH
Europaplatz 3, 69115 Heidelberg, Germany

Printed by Libri Plureos GmbH
in Hamburg, Germany